The Evolution of Life: Teaching, Learning and Training

SCIENCES

Education and Training, Field Director – Jean-Marc Labat

Didactics, Subject Head – Cécile de Hosson

The Evolution of Life: Teaching, Learning and Training

New Approaches on Current Research in the Didactics of Evolution

Coordinated by
Corinne Fortin
Julie Gobert

iSTE

WILEY

First published 2023 in Great Britain and the United States by ISTE Ltd and John Wiley & Sons, Inc.

ISTE Ltd
27-37 St George's Road
London SW19 4EU
UK

www.iste.co.uk

John Wiley & Sons, Inc.
111 River Street
Hoboken, NJ 07030
USA

www.wiley.com

Library of Congress Control Number: 2022948866

British Library Cataloguing-in-Publication Data
A CIP record for this book is available from the British Library
ISBN 978-1-78945-107-8

ERC code:
SH3 The Social World, Diversity, Population
SH3_11 Social aspects of learning, curriculum studies, educational policies

Contents

Introduction

Corinne FORTIN[1] and Julie GOBERT[2]
[1] LDAR Université Paris-Cité, Université de Paris-Est-Créteil, France
[2] ADEF, INSPE Aix-Marseille Université, France

For more than 30 years, numerous research studies in France have been dedicated to the teaching and learning of evolution from elementary school to university. Structured around different theoretical fields, they explore didactic, historical, epistemological and socio-educational issues. This collective work is a contribution to the research dynamics in the didactics of evolution, in the French-speaking context (Quebec, Belgium, France), as well as in connection to other international research to create new research perspectives.

The aim of this book is to put the theoretical, conceptual and methodological references present in the field of the didactics of Earth and Life Sciences into perspective. It analyzes the educational stakes of the teaching and learning of evolution from a historical–epistemological and socio-scientific perspective on teaching and training. This book presents research issues related to students' conceptions, teaching practices, curricular transformations, teacher training, and cultural or religious resistance to the acceptance of evolutionary theory.

The themes dealt with show not only the diversity, but also the liveliness of the research issues surrounding the secularization of knowledge, the epistemological or socio-cognitive obstacles, the problematization of scientific knowledge and the nature of science in evolutionary biology, both from a comprehensive and feasibility perspective.

This book is organized into three parts, each of which contributes to enlightening the other two, owing to their complementary nature.

– The first part, "Teaching of Evolution and Beliefs", is dedicated to the relationship between beliefs (religious, cultural) and the understanding of the theory of evolution. It questions the teaching of evolution by focusing on the difficulties students have in differentiating between the scientific explanatory register and the register of beliefs.

– The second part, "Teaching Evolution: Educational and Training Issues", is focused on the didactic and epistemological contours of scientific knowledge taught in secondary schools, as well as in university teacher training, by integrating historical and contemporary controversies.

– The third part, "Teaching Evolution: Conceptions and Obstacles", focuses on the characterization of difficulties and levers for learning the theory of evolution, and analyzes teaching and learning situations related to the concepts of natural selection, population thinking, contingency and more.

PART 1

Teaching of Evolution and Beliefs

This first part, composed of three chapters, targets religious beliefs and the role of chance in students' acceptance or rejection of the theory of evolution.

The chapter by Hanaa Chalak, Marco Barroca-Paccard and Francis Rouquet revisits the question of conflicts between science and belief among French students in 10th grade (15-year-olds). Based on the reading of historical texts, the authors invite the reader to differentiate between scientific and religious explanatory registers within the framework of "non-overlapping magisteria" and to rework the criteria of demarcation between knowledge and belief for a more operative management of these conflicts. Echoing religious beliefs, the chapter by Magali Coupaud, Julie Gobert, Jérémy Castéra and Alice Delserieys studies students' conceptions of chance in their daily lives and in evolutionary biology. Based on a questionnaire, the study shows that the "probabilistic randomness" conception enables a better understanding of the theory of evolution than the "randomness of fate" conception. The results suggest that there is a plurality of conceptions of chance, discussion of which would facilitate the understanding of evolution. Finally, the chapter by José Luis Wolfs concludes this first part by examining the tensions between science and religious beliefs in their sociocultural dimensions. The author explores the secularized and non-secularized conceptions of science among students in 12th grade[1] in French-speaking Belgium, and shows that a process of self-assignment to a religious community seems to play a decisive role in the acceptance of scientific knowledge in general and the theory of evolution in particular.

---

1. Called *terminale* in the French-speaking context, this is equivalent to 12th grade in the US or year 13 in the UK, with students being 17–18 years of age.

1

Student Positions in Learning about Evolution in Relation to Religious Beliefs and Scientific Knowledge

Hanaa CHALAK[1], Marco BARROCA-PACCARD[2] and Francis ROUQUET[1]
[1] *CREN, Nantes Université, France*
[2] *LirEDD, Haute École Pédagogique Vaud, Switzerland*

1.1. Introduction

In France, the creationist viewpoint seems to be growing. In a 2017 survey[1], it was noted that 31% of respondents, aged between 18 and 24 years old, said that they agreed with the proposition that "God created Man and the Earth less than 10,000 years ago", compared to 14% of those over 65 years old (IFOP 2017). This result may seem surprising since in France the theory of evolution was integrated into Earth and Life Sciences (ELS) lessons during compulsory schooling and most 18–24-year-olds have benefited from it.

The teaching and learning of the theory of evolution encounter difficulties linked, in particular, to the influence of religious beliefs that can sometimes lead students to contest the contents of the theory. Although this questioning of the theory of evolution may seem to be a minority position compared to other countries, such as the United States, it would be wrong to think that this does not constitute an issue for our educational system.

1. Ifop survey on conspiracy, for the Jean-Jaurès Foundation and Conspiracy Watch.

The Evolution of Life: Teaching, Learning and Training,
coordinated by Corinne FORTIN and Julie GOBERT.

One possible explanation is to consider that students have conceptions that correspond to their idea of the world; however, in a psychoanalysis of knowledge, Bachelard (1938/2000) conjectures that primary representations have the status of opinion and are always wrong in principle. Students do not always differentiate between belief, which can be based on "an intuition, a feeling, an intimate conviction, or on the trust or authority granted to a person [...]" (Wolfs 2013, p. 41), and scientific knowledge, which is based:

> on reason, demonstration, proof, that is, on arguments that one must be able to communicate and justify, of which one must be able to account for publicly, so that they can be examined, debated and tested by others, and are therefore in principle independent of the authority status of the person who states them (Wolfs 2013, p. 41).

Several recent studies have focused on this problematic situation for teachers trying to analyze various teaching strategies (Hermann 2008; Hildebrand et al. 2008; Aroua et al. 2012; Fortin 2014; Urgelli et al. 2018). However, the issue of disntinguishing between religious knowledge and beliefs when learning about evolutionnary theory from the students' perspective has been less visited by research. The arrival of the new French high school curricula (M.E.N. 2019) seemed to be a good opportunity to set up a study on this distinction at the curriculum level and from the students' perspective. First, we will present an analysis of the ELS school curricula, and then we will consider the tensions between scientific knowledge and religious beliefs in the context of teaching evolutionary theory and discuss this distinction. We will then look at the relationship between knowledge and beliefs from the point of view of students in the 10th grade by considering, in particular, a situation where they give their opinions on fixist explanations constructed by scientists in the 18th century, before Darwinian theory had been developed.

1.2. A look at French school curricula

Currently, recent French school curricula insist on the need for students to distinguish between knowledge and beliefs. In ELS, for example, one of the skills mentioned in the introduction to the cycle 4 curricula (for students aged 12–15) is that of "distinguishing between what is a belief or an idea and what constitutes scientific knowledge" (M.E.N 2015, p. 342). However, when we look at how the theory of evolution is presented in these curricula, the reference to beliefs, opinions and debates surprisingly does not appear in the presentation of theme 2, "Living Beings and Their Evolution." In the examples of situations and activities, teachers are only asked to call upon science history so that the student can situate the evolution of knowledge in a historical and technical context, without any further details or explanations. We can therefore say that these curricula do not propose a clear operationalization of the differentiation between belief and science. Fortin (2014, p. 72) already said that “since the curricular objectives are based on a requirement of scientific training for students, the management of religious beliefs in the classroom is not taken into account by the curriculum” and this is visibly the case in the ELS curricula in compulsory education.

The new high school curricula (M.E.N. 2019a) mention that it is also necessary to work on this skill in the 10th grade. However, as for cycle 4, it is not remobilized in the presentation of the themes to be addressed in the new high school curricula. Moreover, in the science curriculum for 11th grade, it specifies that one of the main goals of the training is to “understand the nature of scientific knowledge and the methods used to develop it. Scientific knowledge is the result of a rational construction. It is distinct from a belief or an opinion. It is based on the analysis of facts extracted from complex reality or produced during experiments. It seeks to explain reality by material causes” (M.E.N. 2019b, p. 2). Here, we see that knowledge and its characteristics are valued, while what differentiates it from beliefs is not presented. Moreover, high school ELS teachers are asked to “participate in the formation of critical thinking” (MEN 2019c, p. 2). A justification is provided in the preamble: “the exercise of critical thinking is particularly necessary in view of the increasing questioning of the contributions of science”. The development of critical thinking would thus enable students to distinguish between what is science and what is religion.

We can see that the scientific dimension is quite present in the school curricula, but the mention of religious beliefs, in the case of teaching the theory of evolution, is missing, though these school curricula emphasize the distinction between knowledge and beliefs. However, this distinction calls for some vigilance on the part of the teacher, depending on the theoretical framework of problematization (Fabre 2016) in question. The main issue is not only to teach students to distinguish between science and religion, but to develop in them a reflexive and critical posture on science (Mattews 2012), in particular on the complex relationship between science and religion, as we will present in the following sections. Indeed, if the argumentative registers of the religious framework and the scientific framework can be distinguished, the risk is to think that the explanations are, as soon as religion is in question, not scientific. This encourages the religion/science dichotomy. It seems necessary to us for students to know how to distinguish and articulate the language games that interfere in the treatment of the question (Fabre 2016).

1.3. Tensions between scientific knowledge and religious beliefs in the teaching of evolutionary theory

ELS teachers are always confronted with the expression of religious beliefs opposed to the theory of evolution. To remedy this, some try, for example, to set up debates to help students overcoming the opposition between scientific and creationist arguments. Often, in these debates, each person sticks to their position without taking both arguments into account (Fortin 2014). This usually results in low pedagogical effectiveness, probably because, according to Fortin, the management of students' religious beliefs is not taken into account by the curriculum. Other teachers deal with the issue from a solely scientific perspective, without addressing beliefs, and referring to a neutrality of knowledge and the principle of secularism (Urgelli et al. 2018). However, this stance only distances beliefs, and does not always allow students to differentiate their beliefs from scientific knowledge. For example, the expression of certain religious beliefs to challenge the theory of evolution causes tensions that pose many difficulties for teachers.

From the institutional point of view, we can identify changes in position in certain texts published by the Ministry of National Education, which seem to testify to the complexity of the situation. In 2015, the booklet on secularism advocated avoiding confrontations or comparisons between religious discourse and scientific knowledge, as these two registers are not comparable and therefore not opposable. Grjebine and Bouvet (2015) question this recommendation in an article published in the newspaper *Le Monde*[2]:

> What sense would it make to teach Darwinism in schools to children who would hear creationism praised in their family or their church, their temple or their mosque and would invoke it at school? And so, how can we ask a teacher to explain the theory of evolution, without showing that creationism owes more to irrationality than to science?

Avoidance, proposed in 2015, is no longer mentioned in 2020 in the *vade mecum* on "Secularism at school"[3], which recommends remaining firm on the principle that "no question is excluded a priori from scientific and pedagogical questioning (article

2. This article was published on October 26, 2015: https://www.lemonde.fr/idees/article/2015/10/27/l-enseignement-de-la-laicite-doit-maintenir-la-superiorite-de-la-science-sur-la-croyance_4797416_3232.html [Accessed 6 June 2021].

3. This is the *vade mecum*, as updated in October 2020: https://cache.media.eduscol.education.fr/file/Valeurs_republicaines/06/8/202010-Vademecum-Laicite-WEB_1338068.pdf [Accessed 6 June 2021].

12 of the Charter of Secularism in Schools)" (p. 47). This *vade mecum* also specifies that teaching distinguishes between beliefs and knowledge based on rationality: "Through its impartiality and objectivity, it protects the freedom of conscience of pupils and teaches them that certainties are constructed" (ibid.). However, this separation of science from religious beliefs is not a consensus among evolutionists. Dawkins (2006) considers that science must intervene and make a pronouncement on the existence of God: "Either he (god) exists or he doesn't. It is a scientific question; one day we may know the answer and meanwhile we can say something pretty strong about the probability" (ibid., p. 48). Conversely, for Gould (1996), the domains of science and religion should not overlap according to the principle of non-overlapping magisteria (NOMA). This principle of NOMA does not mean that all non-scientific elements should be excluded. It is mainly a matter of separating the explanatory registers in order to signify the difference between science and religion. Lecointre (2009) thus argues in favor of a discussion in science and philosophy classes, within the framework of scientific rationality, social epistemology and beliefs, to differentiate between evolutionary and creationist arguments. From the perspective of educational researchers, Urgelli (2012) defends the idea "that excluding any consideration of the ideological underpinnings of creationism does not allow students to understand the particularities and significance of scientific answers to the question of the origins of humans and life" (ibid. pp. 167–168). We agree with this position and believe that avoidance is not conducive to the development of students' critical thinking. On the contrary, still from the perspective of building school's knowledge, the problematic dimension is essential and the place of criticism, a necessity.

1.4. The compartmentalization of evolutionary knowledge and religious beliefs

There has been, in the scientific literature, a great deal of interest regarding possible rejections of certain concepts in evolutionary theory based on religious beliefs (Carette et al. 2013; Wolfs 2013). This work has shown that religious beliefs can potentially be associated with the rejection of evolutionary theory[4]. However, this does not mean that religious people cannot accept evolutionary theory. Pierre Teilhard de Chardin (1881–1955) was both a recognized evolutionist and a Jesuit priest. He was a scientist when he published his discoveries on the synanthrope, and

4. On the other hand, many more scientists do not exhibit religious beliefs. While only 6% of the American public describes itself as atheist or agnostic, 64% of scientists at "elite" American universities fall into these categories (Ecklund 2010). A study by Larson and Witham (1998) also found that 93% of the members of the National Academy of Sciences are agnostic or atheist, and only 7% believe in God.

a theologian when he problematized evolution in a religious way by centering it on humankind and making the divine the Omega point of evolution[5]. More generally, it is necessary to:

> [...], avoid confusion: not all believers have opposed Darwinian theory and not all those who have rejected it have done so for religious reasons! Many believing biologists, such as Asa Gray, have accepted the Darwinian theory of evolution. And most biologists who rejected this theory did so for scientific reasons, convinced that natural selection was not a sufficient mechanism to explain evolution [...] (Morange 2016, pp. 222–223).

We can thus, by distinguishing religion and science as two separate registers, be an evolutionist and a believer. "A biologist can certainly be a believer, but scientific normativity forbids them from evoking a miracle to solve a biological problem" (Fabre 2016, p. 110).

However, from an individual perspective, consideration of evolutionary theory may challenge religious beliefs. Meadows et al. (2000) asked 17 people who were both practicing Christians and science teachers or science education students[6] about possible conflicts between their personal and professional beliefs. They identify four categories of conflict (Figure 1.1). The first two represent those who intentionally or unintentionally separate their religious beliefs from their learning of evolutionary concepts. The first category is particularly problematic because it may result from a refusal to consider evolutionary theory. Students with fundamentalist beliefs may practice what Cobern (1996) has called "cognitive apartheid". They would use scientific knowledge so as not to jeopardize their academic success, but would keep their religious beliefs under wraps so as to not challenge their beliefs (Meadows et al. 2000). They are able to resolve inherent conflicts by compartmentalizing their conflicting beliefs.

Teachers in the third and fourth categories recognize the conflict between their beliefs about evolution and religion (Figure 1.1). Teachers in the third category began to question their beliefs on creation as they came to a deeper understanding of

---

5. The Omega point represents the ultimate point of development of complexity and consciousness towards which the Universe is moving, whereas the "alpha point" corresponds to its creation, according to Teilhard de Chardin.

6. Participants included two university science teachers, one professor and two graduate students in science education, four high school biology teachers, and eight prospective middle or high school science teachers. All but one were from the south-eastern United States and there were 11 females and six males.

evolutionary concepts that was painful and challenging for them. Teachers in the fourth category chose to combine selected evolutionary concepts with the biblical approach to creation through personal theories, a strategy that helped them to manage conflict. They were comfortable teaching general evolutionary principles, but many remained uncomfortable discussing human evolution.

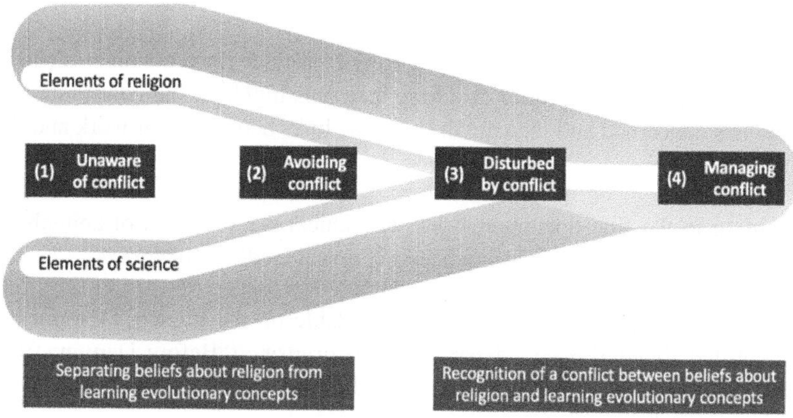

Figure 1.1. *Possible categories of conflict among practicing Christian science teachers or science education students (modified from Meadows et al. (2000, p. 106))*

In order to approach the consideration of beliefs with students and develop a critical reflection on their relationship with science, the history of the progressive construction of the theory of evolution, rich in exchanges between science and religion, seems to be a favorable framework. This is what we tried to apply in two 10th grade classes (with the students being 15–16 years of age), by focusing on the relationship between scientific knowledge and religious beliefs from the students' point of view.

1.5. A case study carried out in two 10th grade classes

We approach the question of the relationship between knowledge and beliefs by specifically studying the productions of students in two 10th grade classes (40 students). They analyzed, individually and then in groups[7], documents

7. The groups formed are heterogeneous in terms of conceptions.

presenting the fixist theory of species[8] in force in the 18th century. We chose the fixist theory so that students could explain why, in their opinion, this theory did or did not seem scientific. Indeed, the Darwinian theory of evolution has already received a lot of media attention and could make students take a position "for" or "against" evolution according to religious background[9]. We chose a scientific theory from the 18th century, now obsolete, for its historical interest, and for the view it gives us of the coexistence of science and religion.

The sheet proposed to the students (Figure 1.2) includes two documents (docs 1 and 3) and refers to a video (doc. 2). These materials highlight the contributions of the naturalists concerned and indicate the possible links between their work and their religious beliefs.

According to the first document, "the hierarchical classification of animals and plants" proposed by Linnaeus was intended to "reveal God's plan".

The second document refers to a video available on the French TV Education website entitled *Avant Darwin: fixisme et transformisme* ("Before Darwin: fixism and transformism"). Students watched only the first part[10] of the video, which presents the fixist theory of species.

The third presents the work of Cuvier who, after studying fossils, showed the disappearance of species in the past and the succession of different faunas in the geological strata. To reconcile his results with the biblical texts (the flood, etc.), he refers to the intervention of catastrophes in order to explain the extinctions he identified, and thus becomes a supporter of catastrophism.

The activity is an opportunity for students to work on the distinction between scientific knowledge and religious beliefs. We would like to point out that the productions studied are the result of a session developed within the framework of a research dissertation[11] conducted by two trainee teachers in the second year of a

8. According to this theory, species are fixed and do not evolve.

9. Students have not previously had a specific course on fixism, but they have surely heard about it in evolutionary theory classes in college.

10. In this part, Pascal Picq, paleoanthropologist and lecturer at the Collège de France, presents the work of Linnaeus and fixism. This video is no longer available online.

11. The dissertation was conducted under the supervision of Francis Rouquet at the École Supérieure du Professorat et des Écoles (ESPE) in Nantes.

master's degree in Teaching, Education and Training Professions[12] on ELS (Alleau and Bugel 2018).

Instruction 1: (individual work) After having seen the documents on **fixism**, a popular theory in the 18th century, explain how the argument developed in these documents is scientific to you. Why?

Document 1 – Carl Linnaeus (1707-1778)
Carl Linnaeus was an 18th century Swedish naturalist, who proposed in his book *Systema naturae*, and again in *Species Plantarum* a hierarchical classification of plants and animals. Species are incorporated into a hierarchy of classifying groups including species, genera, orders, classes and kingdoms. By creating order in the diversity of living beings, Linnaeus sought to unveil the purpose of God and show his greatness. In his opinion, species, created by God, were unchangeable; his conception of living beings therefore being profoundly fixist.

Carl Linnaeus, Systema naturae, Title page of the 1740 edition.© Selva/Leemage

Document 2 – Video : http://education.francetv.fr/matiere/sciences-de-la-vie-et-de-la terre/troisieme/video/avant-darwin-fixisme-et-transformisme

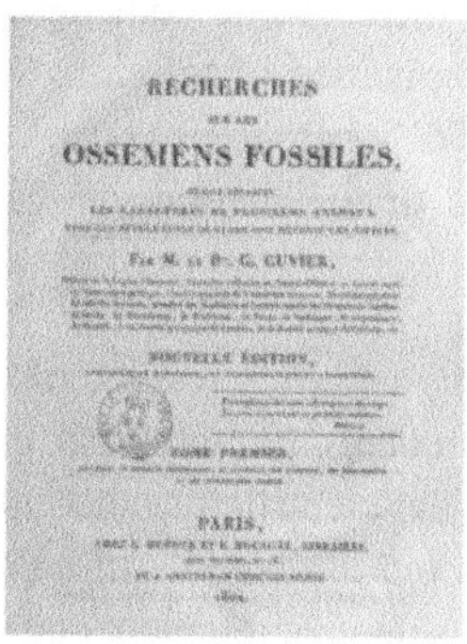

Georges Cuvier, new edition, Paris, G. Dufouret E. d'Ocagne, 1821-1824, tome I. Source : BnF

Document 3 – Georges Cuvier (1769-1832)
"Cuvier showed with his study of fossils that species have died out in the past, and that different fauna follow one another in the geological strata. The desire to reconcile these discoveries with biblical texts led him to become a follower of catastrophism, a fixist theory according to which species became extinct on a mass scale following episodes of catastrophe; devastated regions were then repopulated by flora and fauna from elsewhere. In his life time, Cuvier was furiously opposed to his contemporary Lamarck, who subscribed to the new current of ideas on the transformation of species."

Source of the documents: http://www.reseau-canope.fr/docsciences/Histoire-d-une-pensee-en-evolution.html

Instruction 2: (group work) In groups of three or four, compare your points of view and create a new synthetic opinion (on a separate sheet) on the fixist theory to explain how this theory is scientific or not.

Figure 1.2. *Documents and instructions distributed to 10th grade students (Alleau and Bugel 2018). The first document discusses Carl Linnaeus, the second Georges Cuvier. The original document was in French but has been translated into English for the reader*

12. *Métiers de l'Enseignement, de l'Éducation et de la Formation* (Teaching, Education and Training Professions).

Based on the information available in the documents, we are interested in determining whether students can identify:

– on the one hand, elements of the scientific approach of naturalists at the time (observation, comparison and classification of species, study of fossils and different faunas in geological strata);

– on the other hand, whether they point out the links made with religion as non-scientific (seeking to reveal God's design, reconciling the explanations with biblical texts).

However, it is important to note that the reference to religion in the works of Linnaeus and Cuvier does not obscure the scientific contributions they made at the time, even if these are no longer in agreement with the knowledge accepted today. Moreover, in 1753, Linnaeus established the rules of a binomial nomenclature still used today: "each species is designated by two Latin names, that of the species itself and that of the genus to which it belongs" (Morange 2016, p. 96). The work of Linnaeus and Cuvier presented in the documents should therefore be considered in the context of the time, which was marked by the predominance of religion. What elements do students select from the documents? To what extent can they differentiate between the scientific and religious aspects of the naturalists' work? What place do they give to the links made with religion?

In the following paragraphs, we introduce the qualitative study's results, of both the individual and group responses, produced by the students in two 10th grade classes. We identify whether they define the fixist theory as scientific or not, and we analyze the arguments put forward. The analyses carried out allowed us to identify three categories of position. Some students identify only scientific elements or only religious elements, while others identify and discuss both elements at the same time. We study these positions in detail in the following section, illustrating our remarks with students' statements[13] by highlighting their arguments. We point out that this analysis is based on what the students wrote when they answered the questions asked, and that these statements do not necessarily reflect everything they think about the subject.

13. We present the statements as they were written by the students without spelling corrections.

1.5.1. *Study of students' statements exclusively mobilizing scientific elements*

Seven students mobilize only scientific elements related to the approach taken by the naturalists to position themselves in relation to the fixist theory. They have a dual position with regard to this theory as they think that it is both scientific and not scientific. For Ines, for example, Cuvier's work is scientific because he studied fossils to propose his theory, whereas Linnaeus's work lacks evidence and justification and cannot be considered scientific.

> The argumentation developed in this document *is not scientific because* there is *no proof, no justification*. It is man who has found possible links between living beings in order to classify them into different species. But this is only a proposal and has been contradicted by other theories such as transformism. In doc 3: scientific: *Cuvier made a study of fossils, from that he proposes his own theory*.

Box 1.1. *Ines' statement*

In the same way, Shannon considers Linnaeus's work to be unscientific because he merely classifies living things into groups, whereas Cuvier's work is scientific, because he studies the fauna and flora of the past with reference to catastrophic events that could lead to the distinction of species.

> The argument developed by Carl Linnaeus does not seem scientific because he speaks of a hierarchical classification of classification groups, as there is a periodic classification of chemical elements in physical science, we could believe that it is scientific, *but it is only a classification of species as we could classify anything*.
>
> The Georges Cuvier one seems scientific *because he talks about the past with the fauna and the flora as soon as a species becomes extinct; he talks about a catastrophic event that extinguishes a species but that another one is reborn by its means*.
>
> Finally, the one in the video seems scientific because it says that species should not only be animals that look alike but that, by crossing, form a fertile species (Shannon).

Box 1.2. *Shannon's statement*

Conversely, Sarah identifies both scientific (species classification and transformations) and non-scientific (unchanging species, catastrophes) elements, without a clear reference to God or biblical texts. This reference is implicit in the words she uses to clarify that the content of document 1 is not scientific.

Document 1: the information given to us is not completely scientific because:

Scientific: *hierarchical classification of plants + animals; species transformation*

Unscientific: *unchanging species*; fixed world, fixed species; *catastrophes*

Document 2: the information given to us is scientific: extinct species in the past.

Box 1.3. *Sarah's statement*

We can thus note that these students put forward the elements of the scientific approach of the naturalists and do not question their reference to religion, which challenges us. The content of their statements shows that they give primacy to science, and do not question the place of religious beliefs in the scientific work of the time.

1.5.2. *Study of students' statements that mainly mobilize elements related to religion*

Two students are clearly opposed to the scientific qualification of the fixist theory, because the scientists of the time brought their religious beliefs into their explanations. In this category, the scientific approach of the naturalists is not taken into account in the arguments.

For Louis, for example, the laws created by God cannot be questioned. He points out the link between the immutability of species admitted by Linnaeus, and the impossibility of contradicting religious texts. We can think that this student, even if he does not mention the scientific elements, makes the distinction between religious beliefs and science when he indicates that the laws related to God are not disputed.

Fixism means that there is a relationship with God, the laws were created by God and therefore *we cannot question them*; *we base ourselves on the words of God = not scientific.*

Every fixist theory is linked with religion, always a connection with God.

Based on God, *Carl Linnaeus says that species cannot change even if he is not sure, because we should not contradict religious texts.*

Box 1.4. *Louis' statement*

For Emma, Linnaeus relies on religion and has no proof of what he says. She goes on to explain that, from her perspective, science is not needed to prove anything. Emma's position here is different from the previous one regarding the distinction between science and religious belief. However, as Louis did, her writing does not refer to the elements of the naturalists' scientific approach, which, it should be remembered, cannot be denied despite the intervention of religious beliefs.

> Document 1 does not seem to me to be scientific, since *Carl Linnaeus is based on his religion and he has no proof of what he says*. But, from my point of view, there is *no need for science to prove something.*

Box 1.5. *Emma's statement*

These two students do not choose to call the fixist theory scientific because the work of the naturalists is based on religious beliefs. This reference to religion seems to make them forget about the scientific work done. At first glance, we may think that this position distinguishes science and religious beliefs in a general way. But as we have already seen from the study of the students' productions, this can be different from one student to another.

The group of students, including Emma and Louis, argue that the texts are not scientific because naturalists are influenced by religion and mobilize their religious beliefs. However, they point out that Cuvier conducts research and observations in an attempt to prove that species have become extinct in the past without further clarification.

> We believe that *the texts are not scientific because they contain religious beliefs. These texts are influenced by religion*. But in document 3, we can see that Georges Cuvier tried to research to prove that species became extinct in the past by making observations.

Box 1.6. *Group C's statement*

We can therefore see that, following the group's work, the students still consider the intervention of religious beliefs in the same way. However, Cuvier's work is brought to the fore without further explanation and without allowing a change in their position in favor of the fixist theory.

In this category, the students mainly point to elements relating to religion in order to make a statement about the scientific nature of the fixist theory. The link with religion is not conceived, and the work of the scientists of the time is generally overlooked.

1.5.3. *Study of students' statements that jointly mobilize scientific and religious elements*

Many students (28 students and 6 groups) take into account both scientific and religious elements to make a unique statement about the scientific nature of the theory and justify their position. Indeed, they consider fixism to be scientific overall because the naturalists followed a scientific approach (classifying, observing, etc.), and not because they referred to religion. Evan's statement below illustrates this position.

> Doc 1: scientific*: the diversity of life,* classification
>
> Not scientific: *fixism, created by God because it is an inexplicable theory, which is not proven*
>
> Doc 3: scientific: species become extinct en masse following catastrophic episodes (…) moreover transformation of species geological strata
>
> Not scientific: fixist theory, *biblical text*
>
> Video: Buffon: era that can assemble = be of the same species.

Box 1.7. *Evan's statement*

> Scientific reasoning:
>
> Doc1: observation of species and comparison to classify them. Hierarchization of the groups.
>
> Doc 3: observation of fossils and fauna of different species: comparison
>
> Video: establishes species according to similarities, but observes that the same species can then have very different forms.
>
> Unscientific reasoning:
>
> Doc 1: he uses this reasoning to support his idea that God intended all this. *He mixes belief and science.* According to him, the species are thus "immutable".
>
> Doc 3: he reconciles his idea with his religious beliefs. He becomes a proponent of catastrophism.

Box 1.8. *Eleonore's statement*

These students clearly identify the scientists' approach as scientific and point out the non-scientific link with religion. Some, like Éléonore, go further by mentioning the mixture of science and belief in the work of naturalists, which would explain the explanations used (unchanging species, catastrophism).

The members of group B also specify that religion is not a matter of science, but of beliefs. For them, the principle of the link between scientific observations and biblical texts is in no way a scientific approach.

> The theory of fixism is, on the one hand, scientific. Indeed, it uses scientific approaches such as classification, hierarchization, comparison and observation of concrete elements.
>
> However, on the other hand, it is not scientific because *it involves religion, which is a belief and not a science.* Moreover, Cuvier assumes that there is a link between his observations and the biblical texts, which is not a scientific approach.

Box 1.9. *Group B's statement*

> Scientific fixism or belief?
>
> Fixism is an idea that originated in the 18th century, the Age of Enlightenment, which describes how nature has evolved. For the supporters of fixism, nature, whatever it is, was created by God; each species has been present since the beginning in an unchanging way. The primary purpose of fixism is to demonstrate the greatness of God.
>
> I- An argument based on a belief
>
> Fixism is above all a religious idea and no physical or concrete proof is brought to the defense of fixism; only beliefs are expressed. However, the method operated by fixism, consisting of classifying animals in a hierarchy, is a totally scientific approach, but classified according to a belief, a religious theory. In conclusion, the approach is scientific, but not the information.
>
> II- An idea of the context
>
> *If today, the idea of an evolution of species by a divine force is partially outdated, at the time the feeling and beliefs were not the same, and an argument necessarily implying the presence of God was quite valid.* In the same way that the information observed geologically and involving major biological crises was adequate for the biblical texts.

Box 1.10. *Marius' statement*

Marius' statement (member of group B) draws our attention to the nature of the arguments developed. He admits that the approach followed is scientific despite the intervention of religious beliefs. Unlike some students who separately identify what science and religion are, this student pushes the thinking a little further. He puts fixism in the context of a time when beliefs were not the same and religious arguments were quite valid.

This reflection seems to us to be an interesting one for the students to construct. It corresponds to what we would expect at the end of argumentative and critical work around the tension between scientific knowledge and religious beliefs, because it is strongly argumentative and critical. It would then be appropriate to bring the other students around to such a reflective process.

The writing of group E also places the intervention of religious beliefs in the context of the time when it was unthinkable to contradict biblical texts.

Lucas, a student in this group, also mentioned the same ideas in his individual statement.

> The theory of fixism is based on observations. Whether it is Linnaeus who observes and classifies species, or Cuvier who studies fossils. It is thus based on a scientific approach, since we observe and study facts before deducing something. But it is at the moment of this deduction that religious beliefs intervene. *For, at the time, it would be unthinkable to contradict the biblical texts, for a long time the only sources of knowledge.* The scientists of the time therefore lost the objectivity of their observations with the desire to reconcile their theories with the religious texts. The conclusions drawn from their observations are no longer objective and are influenced by their beliefs. Thus, the theory of fixism is not entirely scientific.

Box 1.11. *Group E's statement*

In these two writings, we can see that the contents are more elaborate and that several elements from the documents are addressed and discussed. This positioning is close to what is expected from the proposed work, where the opposition of scientific and religious registers is overcome by taking into account the historical context in which the analyzed works are situated.

1.6. Discussion of the students' positions

The analysis of the students' statements on the work of the fixist naturalists shows a great diversity of positions on the possibilities of reconciling scientific practice and religious beliefs. The proposed documents explicitly refer to the religious context used by the scientists, which constitutes a way of working on the relationship between science and belief without explicitly asking about the students' personal beliefs. In this sense, the analyses provide us with indicators of the students' ability to differentiate and discuss what science and religious belief are. This didactic strategy is in line with the positions of Lecointre (2009) and Urgelli (2012), who defend the idea that it is necessary to discuss beliefs in science class and avoid excluding them. This would help to develop students' critical thinking skills and to better understand the knowledge constructed by scientists.

In order to decide on the scientific nature of the fixist theory, some students took into account the scientific elements, others the reference to religious beliefs and others did both. These positions can be related to the categories of conflict identified by Meadows et al. (2000) (Figure 1.1), even though they were identified among teachers and students, and in a different context. Indeed, the Meadows et al. model provides information on situations of conflict experienced between a personal religious belief and the scientific knowledge to be taught. In our situation, students who identify only elements of the scientific process or only religious elements could fall into the first category of the Meadows et al. model, in which a separation is made between learning evolutionary concepts and religious beliefs. However, when both scientific and religious elements are identified and discussed, this would correspond to the second and third categories of this model, where the existence of conflicts between religious beliefs and the learning of concepts is recognized. When this opposition is overcome by taking into account the historical context in which the naturalists' work was produced, the fourth category of the model comes into play, corresponding to management of the conflict. The links between our results and the positions of the evolutionists (the NOMA principle of Gould (1996), and the positions of Dawkins (2006)) seem to us to be more complicated to find, given the particularities of students' tasks. In spite of this, we can consider that the reflection of the students who felt that the religious register "has nothing to do" with the scientific one may be in line with the NOMA principle.

The situation proposed to the students thus seems to allow them to find categories in people who may be experiencing a conflict between their beliefs and scientific knowledge. However, it is based on elements of the history of science and does not highlight the personal conflict that the students may experience. It may have been interesting to analyze the students' proposals in terms of their personal beliefs (religion, atheism, etc.). This research also points out why it is interesting to discuss the question of religion in the construction of scientific knowledge. This is why further study is required to investigate the evolution of the students' thinking after the activity and during the sequence. How do their conceptions evolve? How did they take charge of the critical sphere? Moreover, the data collection can be reproduced in another class by conducting interviews with a few students selected from the different categories to try to shed more light on their positions.

1.7. Conclusion

We were interested in the position of 10th grade students, in relation to texts presenting the approaches and work of certain scientists who adhered to the 18th century fixist theory of species. The objective of the proposed activity was to allow students to critically reflect on the relationship between science and religious beliefs in relation to this fixist theory. The qualitative analysis of the individual productions allowed us to show that the students' positions were diverse and varied. These first results show that the distinction between science and religious belief is very delicate for the students. We note that the interweaving of religious considerations and scientific approaches in the work of the 18th-century scientists is not so easy to disentangle. However, the work with this type of text has the merit of allowing difficulties in the demarcation criteria put forward by the students to be identified in order to characterize what is and is not scientific. This can provide guidance for the teacher who wants students to work on the relationship between science and belief in historical explanations. Moreover, it encourages the formation of critical thinking skills, which are emphasized in the new school curricula, so that students can free themselves from academic conformism or imposed neutrality. Moreover, the great diversity of the types of justifications concerning the possibility of whether or not scientific practice and belief can be reconciled clearly shows the potential for a rich critical discussion on this question. This could be done, in our opinion, through a problematization approach (Fabre 2016) that involves holding debates after completing the work, with the aim of pushing the students to question and develop a critical reflection on the place of religion in the age of fixism.

1.8. References

Alleau, C. and Bugel, A. (2018). Une approche épistémologique pour l'enseignement de la théorie de l'évolution. Master's Thesis, ESPE de l'Université de Nantes.

Aroua, S. (2006). Dispositif didactique pour l'enseignement de l'évolution du vivant – Débat en classe pour l'enseignement de la théorie de l'évolution du vivant en Tunisie. PhD Thesis, École Normale Supérieure de Cachan [Online]. Available at: https://tel.archives-ouvertes.fr/tel-00160674 [Accessed 18 September 2018].

Aroua, S., Coquidé, M., Abbes, S. (2012). Controverses dans l'enseignement de l'évolution. Questions de recherches sur les stratégies d'intervention en classe et dans la formation. *RDST – Recherches en didactique des sciences et des technologies*, (5), 4776. doi: 10.4000/rdst.560.

Carette, V. and Poncelet, J.F. (2013). Analyse des difficultés liées à l'enseignement – Apprentissage de la théorie de l'évolution. *Education & Formation*, e-298-03, 87–103.

Cobern, W.W. (1996). Worldview theory and conceptual change in science education. *Science Education*, 80(5), 579–610.

Dawkins, R. (2006). *The God Delusion*. Bantam Books, London.

Ecklund, E.H. (2010). *Science vs. Religion: What Scientists Really Think*. Oxford University Press, New York.

Fabre, M. (2016). *Le sens du problème problématiser à l'école ?* De Boeck, Louvain-la-Neuve.

Fortin, C. (2014). L'enseignement de l'évolution face aux croyances religieuses : quelles perspectives curriculaires possibles ? *Histoire, monde et cultures religieuses*, 32(4), 67–78.

Gould, S.J. (1996). *Full House: The Spread of Excellence from Plato to Darwin*. Harmony Books, New York.

Hermann, R.S. (2008). Evolution as a controversial issue: A review of instructional approaches. *Science & Education*, 17(89), 1011–1032.

Hildebrand, D., Bilica, K., Capps, J. (2008). Addressing controversies in science education: A pragmatic approach to evolution education. *Science & Education*, 17(89), 1033–1052.

IFOP (2017). Le conspirationnisme dans l'opinion publique française [Online]. Available at: https://jean-jaures.org/nos-productions/le-conspirationnisme-dans-l-opinion-publique-francaise [Accessed 15 June 2021].

Larson, E.J. and Witham, L. (1998). Leading scientists still reject God. *Nature*, 394, 313–314.

Lecointre, G., Fortin, C., Guillot, G., Le Louarn-Bonnet, M.-L. (2009). *Guide critique de l'évolution*. Belin, Paris.

Matthews, M.R. (2012). Changing the focus: From nature of science (NOS) to features of science (FOS). In *Advances in Nature of Science Research: Concepts and Methodologies*, Khine, M.S. (ed.). Springer, Dordrecht.

M.E.N. (2015). Programme d'enseignement du cycle des approfondissements (cycle 4). *Bulletin officiel du ministère de l'Éducation nationale*, special n°11, 26 November.

M.E.N. (2019a). Programme de sciences de la vie et de la Terre de seconde générale et technologique. *Bulletin officiel du ministère de l'Éducation nationale*, n°1, 22 January.

M.E.N. (2019b). Programme d'enseignement scientifique de première générale. *Bulletin officiel du ministère de l'Éducation nationale*, n°1, 22 January.

M.E.N. (2019c). Programme de sciences de la vie et de la Terre de première générale. *Bulletin officiel du ministère de l'Éducation nationale*, n°1, 22 January.

Meadows, L., Doster, E., Jackson, D.F. (2000). Managing the conflict between evolution & religion. *The American Biology Teacher*, 62(2), 102–107.

Morange, M. (2016). *Une histoire de la biologie*. Le Seuil, Paris.

Urgelli, B. (2012). Créationnisme et enseignement de l'évolution : quelle éducation citoyenne et laïque ? *ATALA*, 15, 167–181.

Urgelli, B., Guelladress, K., Quentin, A. (2018). Enseigner l'évolution et la nature des sciences face aux contestations d'élèves : essai de modélisation des postures enseignantes. *Recherches en éducation*, 32, 103–116.

Wolfs, J.L. (2013). *Sciences, religions et identités culturelles*. De Boeck Supérieur, Louvain-la-Neuve.

2

Relationships between the Ideas of Randomness and Understanding of the Evolution of Life among French High School Students

Magali COUPAUD, Julie GOBERT, Jérémy CASTÉRA and Alice DELSERIEYS

ADEF, INSPE Aix-Marseille Université, France

2.1. Introduction

The teaching of the evolution of life from elementary school to university is at the heart of many studies around the world (Jensen and Finley 1995; Anderson et al. 2002; Coquidé 2009; Varela et al. 2013; Smith 2010). It is a concept of fundamental importance in modern biology and a complex object of teaching and learning because of the nature of the knowledge and the conceptions students may have of it. Students' difficulty in accepting evolutionary ideas is either directly related to conceptual difficulties intrinsic to evolutionary theory or to non-scientific ideas linked to beliefs fueled by cultural, sociocultural or religious factors (Aroua et al. 2013; Hanley et al. 2014; Yasri and Mancy 2014). In this article, we will focus on the first type of difficulties relating, in particular, to the concepts of randomness and contingency, which are linked to evolutionary mechanisms (Van Dijk and Reydon 2010; Smith 2010). These conceptual difficulties are partly due to the fact that their spatial and/or temporal dimensions make it difficult to perceive them directly

The Evolution of Life: Teaching, Learning and Training,
coordinated by Corinne FORTIN and Julie GOBERT.

(Fiedler et al 2019). Moreover, randomness, underlying evolutionary mechanisms, is a complex and polysemous concept. It takes on different meanings depending on the biological mechanism with which it is associated.

On the students' side, few studies in didactics are interested in students' conceptions of randomness, as a whole. Studies exist in mathematics didactics (Lahanier-Reuter 1999; Coutanson 2010). These are focused on chance from the point of view of randomness. This research, from elementary school to university, is aimed at understanding the teaching of statistics, and is thus restricted (Séré 1993; Briand 2005; Coutanson 2010; Baraké 2011). However, Batanero et al. (2014) state that randomness is a concept to be considered beyond education on statistics and probability. It would also be important in developing students' critical thinking skills. Piaget and Inhelder (1951) captured this interest by conducting experiments with young children. They specified that the appropriation of this notion by children takes place in three stages and that it is only around the age of 11 or 12 years that they are able to really synthesize randomness and take all its dimensions into account. In biology didactics, French studies are interested in elementary school students' conceptions of the evolution of the living world (Jégou-Mairone 2011) or those of middle and high school students (Fortin 2010; Gobert 2014, 2020). To our knowledge, no study considers the conceptions of randomness with middle school students in interaction with their perceptions of the evolution of life.

The objective of our work is to explore, through the development of a questionnaire that we call UnRESt (Understanding Randomness and Evolution for Students), the links between the way students understand the evolution of life and the way they consider randomness, especially in everyday situations. Addressing abstract conceptions of randomness with students to solve learning problems about the evolution of life may be essential in cases where there is a link between understanding evolutionary mechanisms and randomness (Tibell and Harms 2017).

2.2. Multiple meanings of randomness in common thought and epistemological contributions

Beyond biology, randomness is used to explain or predict phenomena. It combines mathematical, physical and philosophical models (Jedrzejewski 2009). Already envisaged by Aristotle, there is still no consensus on its use in the scientific community today. It is a disconcerting topic because it lies at the interface of psychology, cognition, mathematics and epistemology (Lecointre 2016).

In the evolution of life, randomness is a plural concept that refers to three distinct but complementary meanings (Margolis and Laurence 2014), all of which are essential to explaining the mechanisms of the evolution of life (Gayon 2005):

– It can be thought of as luck, an event that occurs unexpectedly in relation to the triggering element. This idea is found in the mechanisms of mutation and exaptation.

– We also speak of a random event, which refers to the laws of probability. The concept of randomness (random probability) is constitutive of the concepts of genetic drift and reproduction in population genetics. Randomness can also refer to the idea of risk, modeled by the laws of probability since Bernouilli (Jedrzejewski 2009). However, depending on their epistemology, probabilities do not necessarily refer to the idea of randomness, as is the case in strict Laplacian determinism (see the relevant chapter in this book), in which probabilities compensate for the ignorance of humankind in the face of objects that are too complex (Merlin 2013). The triggering causes of independent events may be too numerous to be understood, or unobservable by humans. Lesne (2008) reminds us that the great complexity of biological systems is often used by proponents of an "intelligent design", the founder of current species, as an argument that randomness does not exist. According to these supporters, "the living world is too complex to be the result of blind material processes" (Ibid., p. 150).

– In the words of Gould, contingency is defined as: "the tendency of complex systems with many stochastic elements, interacting in multiple ways and in nonlinear ways, to yield theoretically unpredictable results, based on full knowledge of antecedent conditions, but perfectly explainable after the fact" (Gould 2006, p. 70). Indeed, "the weak predictive capacity of the biological sciences is due to the fact that they deal with complex systems, for which we generally have only poor models (thus overwhelmed by the particularities of the cases studied), or for which the initial conditions cannot be determined with exhaustiveness and precision" (Gayon 2006, p. 563). The idea of contingency is extremely important for understanding evolution as a historical science and Jacob tells us that "what separates the evolutionism of Darwin and Wallace from all previous thought is the notion of contingency applied to living beings" (Jacob 1970, p. 170). Very useful but with epistemological weaknesses (Idem., p. 563), Augustin Cournot's definition of randomness-contingency is as follows: a point of encounter between two causal series of independent events (Merlin 2013). Contingency is an essential explanatory element of the mechanism of natural selection. It allows us to understand that genetic mutations occur randomly without being directed by the effect they will have on the individual (Lecointre 2014).

– Merlin (2013) defines "evolutionary" randomness as the absence of design in the sense of Aristotle and the existence of contingency in the sense of Cournot. It allows us to explain the equiprobability of the appearance of favorable, unfavorable or neutral mutations independently of the environment in which they occur. The production of mutations does not depend on a final causality for a better adaptability of a species. "Evolutionary" randomness is thus far removed from the lively debate in the philosophy of science: randomness as determinism or not. In the common sense, chance takes on different meanings: it can be considered a source of luck or bad luck (Gauvrit 2013). A certain number of French people have a finalist vision when it is a question of explaining the evolution of living beings (Bronner 2007). Ordinary thinking finds it difficult to leave room for randomness in the mechanisms of the evolution of living beings. It is cognitively difficult to think that randomness has a strong implication in the correspondence between an improbable mutation that is functional for the individual and the positive effect of this mutation on a better adaptation for the person who carries it. It is easier to think that this correspondence is influenced by a vital force (Ibid. 2007). This idea refers to vitalism, a recurrent theoretical framework in biology strongly criticized by Monod (1970) on the basis of his work in molecular genetics. The "luck" factor is invoked to explain the appearance of a favorable mutation when the number of occurrences that it took for this mutation to appear is not taken into account. Since the evolutionary time is so long, it is sometimes difficult to relate these successes to the number of failures, especially since mutations with a negative effect are very little perceived.

– Randomness can also refer to the idea of destiny, caused by a divine intervention, which refers to a theological vision. According to certain religious beliefs, randomness has no role to play in the course of life. In monotheistic religions, events are determined by a god (Krivine 2016). In the Bible, it is written: "As you do not know the path of the vital breath, how the bones are formed in the womb of the pregnant woman, so you cannot know the work of God, he who makes all things." In the Muslim world, there is a certain fatalistic determinism. The fate of every atom constituting every element is decided by God (Ibid. 2016).

– In addition, in a plural use, randomness can be synonymous with risk, peril. Compared to "hazard" – "randomness" – in French, in English, the word "hazard" is translated as "risk", thus linking randomness and bad luck. In some cases, it is possible to explain events by the capacity, intrinsic or punctual, to be lucky. This idea, uncoupled from any element of religious belief, is not far from a deterministic idea and is in line with the study by Bronner (2007), which indicates that a certain number of French people have a finalist vision of the evolution of life. It refers to a profound epistemology of the evolution of living beings. Evolutionary thinking has not always been as it is today. In scientific language, Krivine (2016) points out that the idea of finalism is still present ("plan of organization", "genetic program") (p. 153). When it comes to the evolution of living things, it is easier to think that the correspondence between mutation and its positive effect on the evolution of the species is more influenced by a vital force than by randomness (Bronner 2007).

Finally, the word contingency, a concept that is very present in the field of biology, is not very present in ordinary everyday language, and we speak rather of randomness when we consider the unexpected character of events or coincidences.

Our interest in students' conceptions of randomness is part of an epistemological reference in which chance could constitute an obstacle (Bachelard 1938) to the understanding of the evolution of living things (Santini 2009). In view of the contradictions of meaning, common or biological, of the word "randomness" highlighted in the epistemological analysis above, we have chosen to study the conceptions of randomness of secondary school students, and the link that can be made with their understanding of the evolution of life. We have therefore chosen, in our study, to create a questionnaire to inform this subject on a large scale, and to measure the link that could exist between their perception of randomness and their understanding of the evolution of life.

2.3. The UnREStquestionnaire: testing the understanding of randomness and the evolution of life among schoolchildren

2.3.1. *Development of the UnRESt questionnaire*

A quantitative approach was chosen to explore the understanding of randomness and the evolution of life among a large number of schoolchildren. The development of the UnRESt questionnaire falls under the umbrella of research on "concept inventories", closed-ended questionnaires originally developed in the United States in the context of physics research in higher education. They now exist in several disciplines (D'Avanzo 2008; Sands et al. 2018), notably for the evolution of life (Anderson et al. 2002; Rutledge and Sadler 2007; Cunningham and Wescott 2009; Yates and Marek 2015). Their construction draws on the different conceptions that research has shown students have of a given concept (Sands et al. 2018) and requires testing based on reliability and validity (Reeves and Marbach-Ad 2016). We used this same methodology for the development of the UnRESt questionnaire. The development is based on qualitative studies already conducted in France around the different conceptions that middle school students have of the evolution of life (Fortin 2010; Gobert 2014; Coupaud 2018), as well as on existing "concept inventories" (Anderson et al. 2002; Rutledge and Sadler 2007; Yates and Marek 2015). The questionnaire was tested several times with different groups of students before its final version was established. During these trials, the validity and reliability of the questionnaire were tested. Validity indicates whether the item being measured, targeted by the questionnaire, is actually being measured. Content validity was tested through critical analysis of the questionnaire by experts on the evolution of life, and construct validity by a cross-analysis of students' responses to the questionnaire and their knowledge. The reliability of a questionnaire indicates whether it provides information on the reproducibility of measurements. Reliability has been estimated by internal consistency measures for items testing the same area of the evolution of life, such as natural selection or randomness (example: contingency). It indicates the extent to which each item contributes to providing information on the same category of test items (DeVon et al. 2007). The statistical tool selected, Cronbach's alpha correlation coefficient, guided the correction of items until all Cronbach's alphas were greater than 0.7 (Coupaud et al. 2019). The goal of UnREst is to understand how middle school students perceive randomness and to explore the links between their perception of chance and how they understand the evolution of life.

2.3.2. *The UnREst questionnaire*

The UnREst questionnaire consisted of 64 closed questions (items) grouped into three sections: (A) items related to the evolution of life; (B) items related to randomness; and (C) personal parameters. In this chapter, only the results of sections A and B will be used. The items in sections A and B are statements for which the students were asked to rate their agreement or disagreement on a four-modality Likert scale.

The analysis a posteriori carried out on the basis of the students' positioning, according to their degree of agreement with these items, makes it possible to group the students' ideas and identify design categories.

In section A, three scientific domains of the evolution of living things were tested by 48 items (the origin of species, such as the common origin of animals; observable evolutionary facts, such as species extinction; and explanatory phenomena, such as natural selection) (Coupaud 2018). The 48 items (24 non-contextualized items (including two duplicated items), in Table 2.1(a), and 24 contextualized items (items associated with a short statement) in Table 2.1(b)) offer both ideas close to those of scientists and alternative ideas (in contrast to the scientific ideas of the evolution of living things) for three different examples: humans (Table 2.1(aA)); present-day animal species (Table 2.1(aB)); and extinct animal species (Table 2.1(aC) and (b)).

			++	+	-	- -	
A	Modern-day humans appeared after the first humans.	Agree					Do not agree
B	The elephants that inhabit the Earth have generally always been the same.	Agree					Do not agree
C	Dinosaurs and crocodiles had a common ancestor.	Agree					Do not agree

a)

27. Today, scientists believe that the wooly mammoth lived millions of years ago. It had thick fur which enabled it to live in a cold climate. In your opinion, how did the wooly mammoth's thick fur coat come about?

			++	+	-	- -	
A	When the climate cooled, mammoths with the thickest fur were warmer and could reproduce better, just like their young, born with thick fur.	Agree					Do not agree
B	The cooling of the environment lead to mammoth being born with thick fur.	Agree					Do not agree
C	Due to the cold, an ancestor of the mammoths grew fur. Then, it gave this thick fur to its young.	Agree					Do not agree
D	It took a long time for mammoths to appear on Earth with thick fur.	Agree					Do not agree
E	Mammoths with thick fur existed because nature can cause these things to happen.	Agree					Do not agree
F	By chance, a mammoth was born with thick fur.	Agree					Do not agree
G	There is no particular reason to explain why mammoths have always had a thick fur. It could have been thinner previously.	Agree					Do not agree
H	Mammoths with thick fur existed because it was their destiny to do so.	Agree					Do not agree

b)

Table 2.1. *Example of items on the evolution of life (section A – (a) three non-contextualized items – (b) eight contextualized items). Translated from French for reference*

Section B includes four meanings of randomness defined from previous exploratory work based on epistemological analysis, and on philosophical classroom discussions that have been the subject of previous analyses[1] (Coupaud et al. 2016):

– "contingency": the meeting of two independent series of random events in reference to the evolutionary sense (Table 2.2(d));

– "probability": related to the random probabilities of occurrence of an event (Table 2.2(a));

– "ignorance": the non-existence of randomness, the result of human's ignorance in reference to Laplace's deterministic vision (Table 2.2(b));

– "Destiny": linked to an external intervention of divine order for the believers or in connection with a vital force, source of chance and/or bad luck (Table 2.2(c)). The two meanings of randomness: random destiny and random chance developed in the previous section have been grouped into a single category in view of the results of philosophical discussions (Coupaud 2018). They highlight that the meanings "destiny" and "chance" are used by students without any real distinction, which may seem paradoxical.

29. You are walking in the street. You meet a friend that you have not seen for a long time. How do you explain this meeting occurring?

			++	+	-	- -	
A	It is a question of probability that you have to know how to calculate.	Agree					Do not agree
B	There must be a reason. Everything that happens to me cannot be explained.	Agree					Do not agree
C	It was by chance that we met, it was destiny.	Agree					Do not agree
D	I had decided to take a certain route, my friend had chosen that day to take the same one when they could have taken another.	Agree					Do not agree

Table 2.2. *Example of randomness items related to a daily situation (section B)*

1. Randomness is a polysemous scientific concept that can generate epistemological and societal difficulties (Coupaud 2018). The philosophical discussions conducted prior to the work to create this questionnaire confirmed that the word "randomness" is not devoid of meaning for the students. They made it possible to collect examples from everyday life and the ideas that the students bring to mind when talking about randomness. This exploratory work feeds the statements of each item of the questionnaire.

The four meanings of randomness selected above were integrated into statements related to three everyday situations relevant to middle school students: meeting a friend on the street (Table 2.2), playing the lottery and rolling dice.

2.4. Data analysis methodology

The UnRESt questionnaire was administered to 1,634 students aged 11–15 from 10 different secondary schools in France during class time with their teacher (mainly an Earth and Life Sciences teacher). The sample is composed of a heterogeneous public, according to the different characteristics of the school: its location (urban or rural); the socioeconomic background of the school (low, medium, high); and the type of school (public, private).

Various methods were used to analyze the data:

– descriptive statistics to expose the general patterns of the 1,634 students. In this chapter, students' conceptions of randomness in everyday situations are presented (Figure 2.1);

– multivariate exploratory analyses to explore students' ideas. They make it possible to manage answers to a large number of questions, a necessary condition given the complexity of the concepts at stake in the evolution of life and randomness. They make it possible to synthesize information by reducing the number of variables, and thus identify trends or test differences in conceptions between groups of individuals, with inter-class analyses associated with Monte Carlo randomization tests (Munoz and Clément 2007) (Figure 2.2). These inter-class analyses allow the generation of groups of individuals according to the responses to items on a variable selected by the researcher, such as the random "probability". The result obtained in the form of a graph can highlight significant differences explained by the other variables in the questionnaire, such as the degree of agreement with natural selection as a mechanism of evolution.

We will first present the way in which the students interviewed consider randomness in examples from everyday life, and we will then present one of the exploratory analyses carried out within the framework of this study, which shows the relationship between the answers to the items concerning the understanding of the evolution of humankind, and the answers to the "probability" randomness items.

2.5. Results and discussion

2.5.1. *Students' conceptions of randomness*

The results of the questionnaire provide an initial descriptive analysis of students' conceptions of randomness (Figure 2.1). The majority of students surveyed strongly or somewhat agreed with the idea that randomness could be found in the form of contingency. Indeed, these students agreed with the idea that randomness was caused by a combination of two sets of independent random events. However, many students (about 40% "strongly agree") also saw destiny as an important aspect to consider in everyday situations where randomness is involved. It seems contradictory that some of the students surveyed both agreed that the occurrence of events was contingent and could be explained by divine intervention causing luck or misfortune.

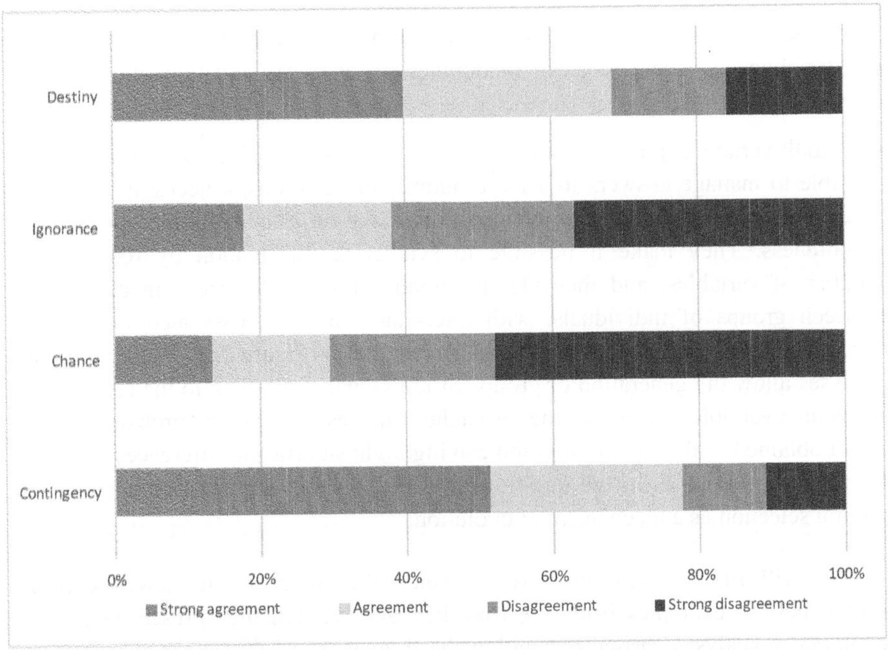

Figure 2.1. *Distribution of junior high students' responses about randomness in relation to examples from everyday life. For a color version of this figure, see www.iste.co.uk/fortin/evolution.zip*

The results also highlight that students rarely recognized randomness as the random probability of an event occurring (about 10% of students fully agreed with the idea of random probability to explain why an event occurred). Similarly, students rarely saw randomness as the current ignorance of humans.

2.5.2. *Influence of students' conceptions of randomness on their understanding of evolution*

Among the different analyses carried out, we present here a cross-class analysis (Figure 2.2), focusing on responses to items related to the understanding of the evolution of humankind, a species that is often difficult to grasp (Jégou-Mairone 2011; Coupaud et al. 2019), and those expressing probabilistic thinking, i.e. randomness perceived as the probability that an event will occur, a form of randomness essential to the understanding of key mechanisms of the evolution of living things. This analysis shows results comparable to other analyses carried out for the case of extinct and current species, namely a relationship between understanding the evolution of life and chance.

In fact, reading the graphs of inter-class analysis allows us to identify more or less distinct groups. The information contained in the graphs of Figures 2.2(b) and 1.3(d) will be related. In Figure 2.2(a), we see that the D1 axis contains almost all of the information (79%). The other axes form background noise due to their low variance. Figure 2.2(d) represents the distribution of students according to their level of acceptance of probabilistic thinking. It shows a clear difference between students on the D1 axis. The numbers, from 4 to 16, represent the centers of gravity of the responses, grouping the students according to their agreement/disagreement with the items related to probabilistic thinking (randomness variable "probability"). Along the D1 axis, there is a clear difference between the group of students expressing strong agreement with probabilistic thinking (score 4) and the groups of students expressing little or no agreement with probabilistic thinking (score 16 or near 16). The Monte Carlo randomization test (Figure 2.2(c)) confirms the significance of this result. The distribution of responses in our sample (observed inertia) is different from that calculated by randomization for 1,000 trials. This means that the differences between the "probabilistic thinking" and "non-probabilistic thinking" groups are quite large, with a margin of error less than 1%. The Monte Carlo randomization test (Figure 2.2(c)) indicates that the difference observed in Figure 2.2(d) is highly significant (p-value <0.001).

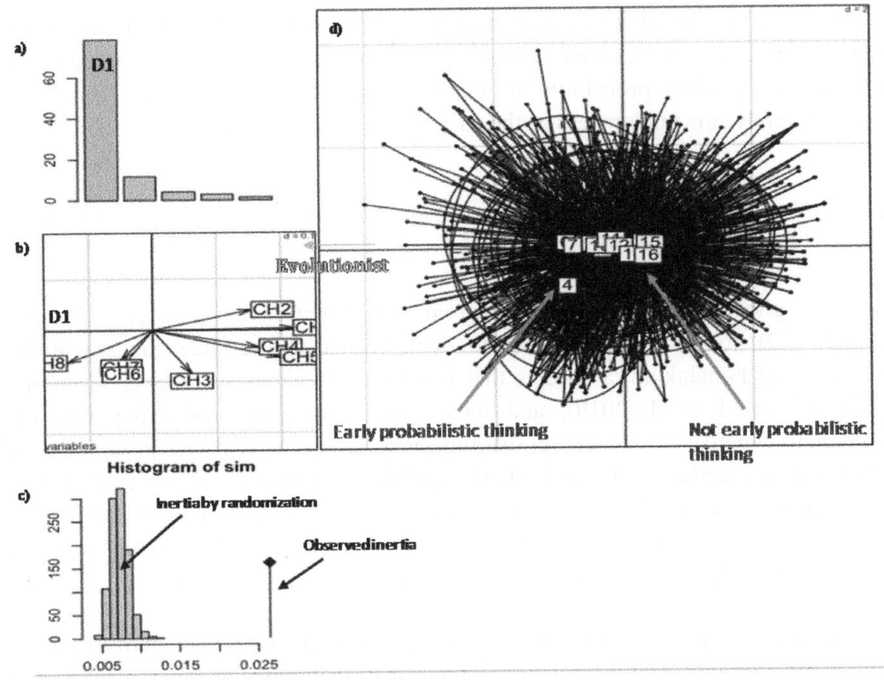

Figure 2.2. *Graphical representations of the inter-class analysis of the items concerning the understanding of human evolution, according to the answers to the "probability" randomness items. (a) Histogram of the eigenvalues of axes D1 and D2. (b) Circle of correlations corresponding to this analysis. (c) Monte Carlo randomization test. (d) Distribution of the "probability" randomness groups according to the D1 and D2 axes (four for agreement with the "probability" items, 16 for disagreement). For a color version of this figure, see www.iste.co.uk/fortin/evolution.zip*

The significance of the D1 axis, necessary for interpretation of the "probabilistic thinking" groups, can be derived from the correlation circle (Figure 2.2(b)). Only three variables (CH1, CH4 and CH5) are more significant (according to the coordinates of these variables on D1) and are related to the statements expressing the degree of agreement or disagreement with natural selection (variable CH1), the role of the environment in the evolution of humankind (variable CH5) and the use of attributes allowing human evolution (variable CH4). We consider agreement with these variables to be linked to evolutionary ideas. The other variables do not provide information that can be analyzed. Students positioned near the top of the vectors identified in Figure 2.2(b) tend to disagree with these variables, giving the legend of the D1 axes on the graph (Figure 2.2(d)). In other words, the direction of the D1 axis

(Figure 2.2(d)), obtained by the circle of correlations (Figure 2.2(b)), is oriented in such a way that, on the right side, we have the students who disagree with the variables CH1, CH5 and CH4, and who therefore reject the evolutionary ideas. It can thus be said that the differences between the groups (4–16) identified in Figure 2.2(d) relate to evolutionary explanations, such as the mechanism of natural selection, adaptation by the use of attributes or adaptation through a change in the environment. In other words, students who perceive randomness in everyday life as a random event (groups in Figure 2.2(d)) also share an evolutionary conception for the human species (variables CH1, CH2 and CH5 in Figure 2.2(b)) and vice versa. From the results of this analysis, it can be concluded that students' conceptions of randomness perceived as a probability of an event occurring and their understanding of human evolution are related.

Other analyses, such as orthogonal principal component analyses on instrumental variables, have shown that this relationship between the understanding of the evolution of life and the understanding of randomness does not depend on the student's grade level, nor on their degree of religious belief. We performed the same type of analysis for the four ideas of randomness proposed in this chapter (contingency, probability, ignorance in the face of a complex situation, consequence of destiny). It emerges from these analyses that the notions of "probability" randomness and "destiny" randomness have a particular link to the students' responses on their understanding of the evolution of life. Indeed, whatever their class level, the students in favor of the idea of "probability" randomness:

– explained the evolution of extinct species by invoking a long period of time and changes in environment;

– for present-day species, were more in agreement with evolutionary explanations with a Lamarckian tendency (items CA5: "It was the appearance of food high up in the trees that caused the birth of long-necked giraffes."; CA4: "As a giraffokeryx pulled on its neck to catch the food in the trees, its neck became longer. This giraffokeryx then passed on this long neck to its young.");

– disagreed that there was no explanation for the evolution of extinct species and current species, and disagree with non-evolutionary explanations (item CA8: "There is no particular reason why giraffes have a long neck. It could have been shorter.").

Furthermore, regardless of their level of religious belief, junior high students who favor the idea of "destiny" randomness tend to choose theological explanations for the evolution of species, such as:

– for humankind, items CH6 ("Men exist today with long legs because it was their destiny"), CH7 ("Men exist today with long legs because nature does things right") and CH8 ("There is no particular reason why men have long legs. They could have been shorter");

– for extinct species, items CD6 ("Mammoths with thick fur existed because it was their destiny"), CD7 ("Mammoths with thick fur existed because nature does things right") and CD8 ("There is no particular reason why mammoths had thick fur. It could have been less thick");

– for the current species, items CA6 ("Giraffes exist today with a long neck because it was their destiny"), CA7 ("Giraffes exist today with a long neck because nature does things right") and CA8 ("There is no particular reason why giraffes have a long neck. It could have been shorter.").

Furthermore, analyses revealed no significant relationship between junior high students' responses to the contingency randomness, ignorance randomness and the evolutionary understanding items, regardless of the species presented.

2.5.3. *Various conceptions of randomness*

The results of this study show that the word "randomness" is not devoid of meaning for the middle school students surveyed. However, the different meanings chosen by the students show that this word can be a source of confusion, as pointed out by Chrétien and Gaud (1998). The study reveals that more than half of the middle school students surveyed are in favor of randomness, synonymous with contingency, as well as with randomness, synonymous with luck, misfortune or destiny, and more than a third of the middle school students surveyed are in favor of the idea of probabilitý, as well as with the idea that randomness does not exist.

2.5.4. *A lack of relationship between "contingency" randomness and the degree of understanding of the evolution of life*

Contingency is a key explanatory element of the mechanism of natural selection (Lecointre 2014b). The results obtained in our study reveal no connection between the idea of "contingency" randomness and the understanding of the evolution of living things for the middle school students surveyed. However, more than half support this idea of rather objective randomness in everyday contexts. Other qualitative analyses also reveal that few students perceive the idea of contingency (Coupaud 2018). The middle school students interviewed may not realize the full complexity of this idea of randomness. Students mobilizing this conception of randomness for everyday contexts do not use this conception when explaining the evolution of living things. This idea of "contingency" randomness is very little approached in high school. It is little discussed in high school teaching curricula. It is only noted in the teaching of visual arts in a particular context, that of the encounter of accidents or opportunities (MENESR 2016a). However, this context can also refer to the idea of randomness, the result of providence. Integrating this idea of contingency into the teaching of different disciplines could allow students to extend this idea to contexts other than those of everyday life, particularly in science classes, and thus allow them to access evolutionary thinking.

2.5.5. *A relationship between "probability" randomness and the degree of understanding of the evolution of life, according to the context of the items*

The results of the study reveal that the consideration of certain ideas of randomness, such as that of randomness calculable by probabilities, by the junior high students surveyed, are related to their level of understanding of the evolution of life. For all the species presented, the idea of "probability" randomness influences the choice of the junior high students questioned in their answers on their understanding of the evolution of life. When the students choose randomness, as calculable by probabilities, they are more favorable to evolutionary explanations.

Probabilities allow us to model randomness and define possible events. They participate in the prediction of future events, as well as in the analysis of past events. This idea of randomness is an essential element for the understanding of biological mechanisms linked to evolution, such as the reproduction of individuals or genetic drift. It allows us to model the possibilities of genetic distribution at the time of gamete formation or fertilization, and to understand the variability of individuals within the same species. The results of our study show that students who support the idea of "probability" randomness in a real-life context have more evolutionary thinking than others. We can think that students who are able to mobilize this idea of "probability" randomness have an easier time understanding the transmission of traits with modifications in successive generations. This idea would allow them to understand that genetic events occur randomly without being directed. It would also allow them to understand why it is difficult to accurately predict the composition of the genotype of an individual or the genetic frequencies within successive generations of populations. The teaching of probability should be broader in scope and not restricted to the teaching of mathematics. It would provide access to evolutionary thinking. In the study conducted in Chapter 7 of this book, Laurent Theis and Julie Gobert identify obstacles to the teaching of the ideas of randomness in the theory of evolution, in the context of an interdisciplinary teaching situation in mathematics and Earth and Life Sciences.

2.5.6. *A relationship between "destiny" randomness and the degree of understanding of the evolution of life*

The results of this study show that the idea of subjective randomness, the result of a destiny, influences the degree of understanding of the evolution of life among the college students surveyed. When middle school students choose this idea of randomness, they seem to be less in agreement with evolutionary thinking. They choose non-rational or non-evolutionary explanations. These results hold true regardless of the species studied or the religious beliefs of the students surveyed. It is not surprising to measure a relationship between randomness and understanding of the evolution of life, but quite surprising to see these results hold after removing the effect of religious beliefs. This result suggests that the link between this conception of chance and students' understanding of the evolution of life goes beyond belief in god as it is resistant to the removal of the effect of religious beliefs. We can speculate that students may choose this idea of "destiny" randomness for two reasons: either for religious reasons or for non-theological reasons.

2.6. Conclusion

We wish to emphasize here the complexity of elements of evolutionary thought that should be clarified with the students beyond a discussion of religious beliefs. Elements, other than religious ones, could be a trigger for debate or discussion within the class, thus detaching the discourse from religious positions. It seems possible to have religious convictions and to accept randomness, as long as we do not focus on religious beliefs but on the understanding of science (Reiss 2005). It is therefore not a question of believing in randomness, but of understanding to what extent the proposed models of randomness are useful in explaining evolutionary facts. These exchanges around different ideas of randomness would also avoid reinforcing the science-religion dichotomy that authors such as Jones and Reiss (2007) show is uninteresting or even harmful to students' engagement with science. The word "destiny" could be a lever in discussions to talk about beliefs without questioning students' religion.

A diversified teaching of the scientific concept of randomness could also allow students to access explanations other than theological ones, thereby limiting their rejection of the evolution of living beings. These results open up avenues of reflection concerning the content of curricula, as well as teaching practices.

The results of this study allow us to consider implications for teaching practices, such as reinforcing the scientific concept of "randomness" in curricula by expanding its teaching to disciplines other than mathematics. Examples of incorporating randomness into disciplines other than mathematics are found in some studies, such as those related to the nature of science (Stavrou et al. 2014).

2.7. Appendix

			++	+	-	--	
A	Les Hommes actuels sont apparus après les premiers Hommes.	D'accord					Pas d'accord
B	Les éléphants qui peuplent la Terre ont globalement été toujours les mêmes.	D'accord					Pas d'accord
C	Les dinosaures et les crocodiles ont eu un ancêtre commun.	D'accord					Pas d'accord

a)

27. De nos jours, les scientifiques pensent que le mammouth vivait il y a des millions d'années. Il possédait une fourrure épaisse qu'il leur permettait de vivre dans un climat froid. D'après toi, comment la fourrure épaisse des mammouths est-elle apparue ?

			++	+	-	--	
A	Lorsque le climat s'est refroidi, les mammouths avec la fourrure la plus épaisse ont eu plus chaud et ont pu mieux se reproduire, tout comme leurs petits, nés avec une fourrure épaisse.	D'accord					Pas d'accord
B	C'est le refroidissement de l'environnement qui a provoqué la naissance de mammouths à fourrure épaisse.	D'accord					Pas d'accord
C	A cause du froid, un ancêtre du mammouth a laissé pousser sa fourrure. Puis il a transmis cette fourrure épaisse à ses enfants.	D'accord					Pas d'accord
D	Il a fallu un temps très long pour que les mammouths apparaissent sur Terre avec une fourrure épaisse.	D'accord					Pas d'accord
E	Les mammouths avec une fourrure épaisse ont existé car la nature fait bien les choses.	D'accord					Pas d'accord
F	Par hasard, un mammouth est né avec une fourrure plus épaisse.	D'accord					Pas d'accord
G	Il n'y a aucune raison particulière qui explique que les mammouths aient eu une fourrure épaisse. Elle aurait pu être moins épaisse.	D'accord					Pas d'accord
H	Les mammouths avec une fourrure épaisse ont existé car c'était leur destinée.	D'accord					Pas d'accord

b)

Table 2.3. *French version of Table 2.1*

29. Tu te promènes dans la rue. Tu rencontres un copain (ou une copine) que tu n'as pas vu depuis longtemps. Comment expliques-tu cette rencontre ?

			++	+	-	--	
A	C'est une question de probabilité qu'il suffit de savoir calculer.	D'accord					Pas d'accord
B	Il doit y avoir une raison. Tout ce qui m'arrive est explicable.	D'accord					Pas d'accord
C	J'ai eu de la chance de la rencontrer, c'était ma destinée.	D'accord					Pas d'accord
D	J'ai choisi de passer par un chemin et justement, mon copain (ou ma copine) a choisi ce jour-là de passer par le même chemin alors qu'il (elle) aurait pu en prendre un autre.	D'accord					Pas d'accord

Table 2.4. *French version of Table 2.2*

2.8. References

Anderson, D.L., Fisher, K.M., Norman, G.J. (2002). Development and evaluation of the conceptual inventory of natural selection. *Journal of Research of Science Teaching*, 39(10), 952–978.

Aroua, S., Coquidé, M., Abbes, S. (2013). Enseigner l'évolution du vivant dans un contexte concordiste. *Review of Science, Mathematics et ICT Education*, 7(1), 5–26.

Bachelard, G. (1938/2000). *La formation de l'esprit scientifique : contribution à une psychanalyse de la connaissance objective*. Vrin, Paris.

Baraké, F. (2011). Autour des conceptions du hasard et des probabilités élémentaires chez les jeunes déficients intellectuels légers. Education science PhD Thesis, Université Paris 5.

Batanero, C., Arteaga, P., Serrano, L., Ruiz, B. (2014). Prospective primary school teachers' perception of randomness. In *Probabilistic Thinking*, Chernoff, E.J. and Sriraman, B. (eds). Springer, Dordrecht.

Briand, J. (2005). Une expérience statistique et une première approche des lois du hasard au lycée par une confrontation avec une machine simple. *Recherches en didactique des mathematiques*, 25(2), 247–282.

Bronner, G. (2007). La résistance au darwinisme : croyances et raisonnements. *Revue française de sociologie*, 3(48), 587–607.

Chrétien, C. and Gaud, D. (1998). Qu'est-ce le hasard ? Comment le mathématiser ? *Repères- IREM*, 32, 81–110.

Coquidé, M. (2009). Compréhension et repérage des conceptions des élèves de primaire, de collège et de lycée concernant l'évolution des vivants. In *L'évolution du vivant, un enseignement à risque ?*, Coquidé, M. and Tirard, S. (eds). Vuibert, Paris.

Coupaud, M. (2018). Évolution du vivant et hasard : étude quantitative des conceptions d'élèves de collège dans le contexte français. Education science PhD thesis, Université Aix-Marseille.

Coupaud, M., Delserieys, A., Jégou, C., Agostini, M., Brandt-Pomares, P. (2016). Idées des élèves sur le hasard lors de discussions à visée philosophique. *Educational Journal of the University of Patras UNESCO Chair*, 3(2), 340–349.

Coupaud, M., Castéra, J., Cheneval-Armand, H., Brandt-Pomares, P., Delserieys, A. (2019). Développer un questionnaire pour étudier les conceptions de l'évolution du vivant d'élèves de collège : entre cadres didactique et psychométrique, *RDST*, 20, 27–59.

Coutanson, B. (2010). La question de l'éducation statistique et de la formation de l'esprit statistique à l'école primaire en France. Étude exploratoire de quelques caractéristiques de situations inductrices d'un enseignement de la statistique au cycle III. PhD Thesis, Université Lumière-Lyon II.

Cunningham, D.L. and Wescott, D.J. (2009). Still more "Fancy" and "Myth" than "Fact" in students' conceptions of evolution. *Evolution: Education and Outreach*, 2, 505–517.

D'Avanzo, C. (2008). Biology concept inventories: Overview, status, and next steps. *BioScience*, 58(11), 1079–1085.

DeVon, H.A., Block, M.E., Moyle-Wright, P., Ernst, D.M., Hayden, S.J., Lazzara, D.J., Savoy, S.M., Kostas-Polston, E. (2007). A psychometric toolbox for testing validity and reliability. *Journal of Nursing Scholarship*, 39(2), 155–164.

Fiedler, D., Sbeglia, G.C., Nehm, R.H., Harms, U. (2019). How strongly does statistical reasoning influence knowledge and acceptance of evolution? *Journal of Research in Science Teaching (JRST)*, 56(9), 1183–1206.

Fortin, C. (2010). *L'évolution à l'école : créationnisme contre darwinisme ?* Armand Colin, Paris.

Gauvrit, N. (2013). Conceptions du hasard et biais probabilistes chez des enseignants du second degré : effet d'une formation courte. *Statistique et enseignement*, 42(2), 53–66.

Gayon, J. (2005). Évolution et hasard. *Laval théologique et philosophique*, 61(3), 527–537.

Gobert, J. (2014). Processus d'enseignement-apprentissage de raisonnements néodarwiniens en classe de sciences de la Vie et de la Terre. Education Science PhD Thesis, Université de Caen Basse-Normandie.

Gould, S.-J. (2002). *The Structure of Evolutionary Theory*. The Belknap Press of Harvard University Press, Cambridge, MA.

Hanley, P., Bennet, J., Ratcliffe, M. (2014). The inter-relationship of science and religion: A typology of engagement. *International Journal of Science Education*, 36(7), 1210–1229.

Jedrzejewski, F. (2009). *Modèles aléatoires et physique probabiliste*. Springer Science, Paris.

Jégou-Mairone, C. (2011). Des élèves de 9–12 ans de l'école primaire française et l'évolution des espèces vivantes. *Review of Science, Mathematics And ICT Education*, 5(1), 81–96.

Jones, L. and Reiss, M.J (2007). *Teaching About Scientific Origins: Taking Account of Creationism*. Peter Lang, New York.

Krivine, H. (2016). *Petit traité de hasardologie*. Cassini, Paris.

Lahanier-Reuter, D. (1999). *Conceptions du hasard et enseignement des probabilités et statistiques*. PUF, Paris.

Lecointre, G. (2014). *L'évolution. Question d'actualité ?* Éditions Quae, Versailles.

Lecointre, G. (2016). Postface. In *Petit traité de hasardologie*, Krivine, H. (ed.). Cassini, Paris.

Lesne, A. (2008). Complexité du vivant, sélection naturelle et évolution. *Natures Sciences Sociétés*, 16(2), 150–153.

Margolis, E. and Laurence, S. (2014). Concepts. In *The Stanford Encyclopedia of Philosophy*, Zalta, E.N. (ed.). Stanford, Spring Edition.

MENESR (2016a). Programme d'enseignement du cycle des apprentissages fondamentaux, du cycle de consolidation et du cycle des approfondissements [Online]. Available at: http://www.education.gouv.fr/pid285/bulletin_officiel.html?cid_bo=94708 [Accessed 9 April 2018].

Merlin, F. (2013). *Mutations et aléas : le hasard dans la théorie de l'évolution*. Hermann, Paris.

Monod, J. (1970). *Le hasard et la nécessité*. Le Seuil, Paris.

Munoz F. and Clément P. (2007). Des méthodes statistiques originales pour analyser les conceptions d'enseignants de plusieurs pays à partir d'un questionnaire sur des questions vives. *Actes du Congrès AREF 2007 – Actualité de la Recherche en Education et en Formation*, Strasbourg.

Piaget, J. and Inhelder, B. (1951). *La genèse de l'idée de hasard chez l'enfant*. Presses Universitaires de France, Paris.

Reeves D. and Marbach-Ad, G. (2016). Contemporary test validity in theory and practice: A primer for discipline-based education. *CBE – Life Sciences Education*, 15, 1–9.

Reiss, M.J. (2005) The importance of affect in science education. In *The Affective Dimensions of Cognition: Studies from Education in the Sciences*, Alsop, S. (ed.). Kluwer, Dordrecht.

Rutledge, M.L. and Sadler, K.C. (2007). Reliability of the measure of acceptance of the theory of evolution (MATE) instrument with university students. *The American Biology Teacher*, 69(9), 332–335.

Sands, D., Parker, M., Hedgeland, H., Jordan, S., Galloway, R. (2018). Using concept inventories to measure understanding. *Higher Education Pedagogies*, 3(1), 173–182.

Santini, J. (2009). Les conceptions des élèves : des données empiriques aux résultats inférés. Questionner l'implicite d'un objet des didactiques à partir des conceptions du mécanisme sismique. In *Questionner l'implicite. Les méthodes de recherche en didactiques*, Cohen-Azria, C. and Sayac, N. (eds). Presses Universitaires du Septentrion, Villeneuve d'Ascq.

Séré, M. (1993). Le déterminisme et le hasard dans la tête des élèves. *Bulletin de l'Union des Physiciens*, 740, 87–96.

Smith, M.U. (2010). Current status of research in teaching and learning evolution II: Pedagogical issues. *Science & Education*, 19(6–8), 523–538.

Stavrou, D. and Duit, R. (2014). Teaching and learning the interplay between chance and determinism in nonlinear system. *International Journal of Science Education*, 36(3), 506–530.

Tibell, L.A.E. and Harms, U. (2017). Biological principles and threshold concepts for understanding natural selection – Implications for developing visualizations as a pedagogic tool. *Sciences & Education*, 26, 953–973

Van Dijk, E.M. and Reydon, T.A.C. (2010). A conceptual analysis of evolutionary theory for teacher education. *Science & Education*, 19, 6–7.

Varela, M.A.C., Dos Santos, I.B.C., Ferreira J.H.B.P., Bussab, V.S.R. (2013). Misunderstandings in applying evolution to human mind and behavior and its causes: A systematic review. *EvoS Journal: The Journal of the Evolutionary Studies Consortium*, 5(1), 81–107.

Yasri, P. and Mancy, R. (2014). Understanding student approaches to learning evolution in the context of their perceptions of the relationship between science and religion. *International Journal of Science Education*, 36(1), 24–45.

Yates, T.B. and Marek, E.A. (2015). A study identifying biological evolution-related misconceptions held by Prebiology high school students. *Creative Education*, 6, 811–834.

3

The Theory of Evolution: Possible Tensions between Science and Religious Beliefs

José Luis WOLFS
Université Libre de Bruxelles, Brussels, Belgium

3.1. Introduction

The sources of difficulties related to the teaching and learning of the theory of evolution are numerous and varied: the very complexity of evolutionary biology, a broader misunderstanding of the scientific process, or forms of confusion or interference between scientific knowledge and personal beliefs, in particular religious beliefs[1]. This is evidenced by numerous research studies on the total or partial rejection of the theory of evolution in the name of religious beliefs by students (Fortin 2009, Perbal et al. 2006; Aroua et al. 2012) or by teachers (e.g. Clément 2014). In this chapter, we will focus specifically on the possible tensions between science and religious beliefs, first in general and then in the case of teaching evolutionary theory.

---

1. Personal beliefs that may interfere with a Darwinian conception of evolution are obviously not only religious. Guillot (2015), for example, notes the strong presence among French high school students of non-religious conceptions of the finalist type (evolution would be "naturally" oriented towards the production of increasingly complex species adapted to their environment, without explicitly taking into account the following two explanatory mechanisms: (1) the role of chance during mutations and (2) the role of natural selection.

The Evolution of Life: Teaching, Learning and Training,
coordinated by Corinne FORTIN and Julie GOBERT.

Here is how Weber summarized the characteristics of a secularized conception of science and the consequences he drew from it:

> Science (…) has no knowledge of "miracles" nor "revelation". (…) the course of things must be explained without the intervention of any of those supernatural elements to which empirical explanation denies any causal character; it cannot be explained otherwise than by the method which science endeavors to apply. And the believer can admit this without any infidelity to his faith (Weber 1919, pp. 104–105).

In fact, however, the consequence that Weber draws from this, "and the believer can admit this without any infidelity to his faith", has turned out to be less simple than it first appears. The development of a "modern" and "secularized" conception of science has indeed tended to produce complex, numerous and contradictory effects with regard to the relationship to religion. As Lambert notes:

> It is trite to recall that science could (and still can) generate atheism (scientism, materialism), overall reinterpretations (demythologization, critical exegesis), fundamentalist reactions (creationism) or innovations (Christian Science, Church of Scientology, New Age, parasciences, etc.) (Lambert 2000, p. 102).

For this reason, we feel it is important to examine how students experience the possible tensions between science and religious beliefs, before focusing on the case of evolutionary theory.

The approach taken in this chapter is thus to examine:

1) the theoretically possible positions between science and religious beliefs (considered in a general way, rather than in the case of evolutionary theory);

2) the extent to which high school seniors in 16 countries have a secularized view of science;

3) the factors likely to influence both the development of a secularized conception of science and the degree of knowledge and/or acceptance of the theory of evolution;

4) finally, some pedagogical and didactic avenues to help students to differentiate between scientific and religious discourses.

3.2. Theoretically possible positions between science and religious beliefs

Various historical and sociological studies (e.g. Minois 1990; Urvoy 2006; El Asri 2009; Leru 2010; Chaline and Grimoult 2011; Portier et al. 2011), several existing reading grids (e.g. Lambert 1999; Rasi 2003) and data from field surveys (e.g. Chabchoub 2001; Martin-Hanssen 2008) have led us to develop a model aimed at taking into account different theoretically possible positions between science and religious beliefs (Wolfs 2013, 2018; Wolfs et al. 2021).

In the educational field, the scientific literature also mentions several typologies aimed at listing different possible positions between science and religious beliefs (e.g. Barbour 1990; Haught 1995; Stenmark 2004; Alexander 2007; Taber et al. 2011; Billingsley et al. 2013; Yasri et al. 2013; Yasri and Mancy 2014; O'Brien and Noy 2015). Most of these typologies consider between three and five main positions:

– one is the potential conflict between science and religious beliefs;

– a second, which constitutes an alternative to the conflict, is that of the independence between science and religious belief. In this case, science and religion are considered to be such fundamentally different domains (in terms of the subjects tackled, the questions asked, the methods chosen, etc.) that there is therefore no need for opposition;

– a third position expresses the idea of a complementarity that is deemed possible, even desirable, between science and religious belief;

– a fourth position, which is sometimes difficult to distinguish from the previous one, is that of rapprochement leading to a form of confusion between science and religious beliefs;

– finally, some authors mention the case of individuals who do not have a clear opinion on the question, either because they have never explored it or because they consider it undecidable or are constantly re-evaluating the contribution of science and religion to their understanding of the world.

The model we will refer to in this chapter (Wolfs 2013, 2018; Wolfs et al. 2021) includes seven contrasting positions, defined as ideal types[2]:

1) the rejection of science in the name of fideist conceptions, most often taking the form of a literalist reading of the Scriptures;

2) classic concordism establishing, for example, direct comparisons between passages of sacred Scripture and scientific data;

2. Yasri et al. (2013) also propose a model with seven main possible positions between science and religious beliefs. These are, however, defined in significantly different ways. A systematic comparison between this model and ours, with respect to their theoretical foundations, their methods of construction and the categorizations obtained, is proposed in Wolfs et al. (2021).

3) reversed concordism seeking to find God through an individual vision of science (in the form of a "great architect" or "intelligent design");

4) the autonomy of science in relation to religious beliefs, with forms of complementarity between science and religious beliefs other than concordism (e.g. Gould 2000);

5) the autonomy of science in relation to religious beliefs without seeking complementarity;

6) rationalist criticism, in the name of science, of religious beliefs (non-scientist form);

7) rationalist criticism (scientist form).

Table 3.1 provides a better visualization of these different conceptions, as well as the underlying distinguishing criteria[3]. These are defined in a much more developed, nuanced and argued manner in Wolfs (2013, 2018) and Wolfs et al. (2021).

<table>
<tr><th>Position 1</th><th>Position 2</th><th>Position 3</th><th>Position 4</th><th>Position 5</th><th>Position 6</th><th>Position 7</th></tr>
<tr><td rowspan="2">Rejection of science, in the name of fidest conceptions</td><td rowspan="2">Classic concordism</td><td rowspan="2">Reversed concordism</td><td colspan="2">Mutual autonomy between science and religious beliefs</td><td colspan="2">Rationalist criticisms, in the name of science, of religious conceptions</td></tr>
<tr><td>with a search for a form of non-concordist complementarity</td><td>without seeking complementarity</td><td>Non-scientist form</td><td>Scientist form</td></tr>
<tr><td colspan="3">Pre-eminence of religious beliefs</td><td colspan="2">Lack of pre-eminence between science and religious beliefs</td><td colspan="2">Pre-eminence of science</td></tr>
<tr><td>No "alliance" research</td><td colspan="3">Search for the "alliance" between science and religion, from very different perspectives</td><td colspan="3">No search for a science–religion "alliance"</td></tr>
<tr><td colspan="3">Confusion between science and religion</td><td colspan="3">Incommensurability between the scientific explanatory register and religious beliefs</td><td>Confusion between science and religion</td></tr>
</table>

Table 3.1. *Types of positions between science and religious beliefs*

3. The same person can possibly adhere to several of these positions, depending on the themes addressed. Positions 1–3 and 7 lead to situations of conflict, confusion or amalgam between science and religions. Positions 4–6 lead to a delimitation of their respective regimes and to different modes of cohabitation.

3.3. Conceptions of the secularization of science among students from 16 countries

Based on this theoretical framework, a questionnaire was designed and administered to more than 6,000 students in senior high school classes in 16 countries to date[4]. This was a significantly revised version of the questionnaire published in 2014 (Wolfs et al. 2014). A new version has been published since, integrating in particular the "scientism" position that was absent in the first version (Wolfs et al. 2022). It consisted of about 30 closed questions accompanied each time by a six-point Likert scale, ranging from "strongly disagree" to "strongly agree". For each of the positions envisioned by the baseline model, the student scored an average between -10 and +10[5].

To characterize a "secularized" conception of science, two indicators[6] were defined: the rejection of fideist and concordist positions (in classic and reversed forms) and the adherence to the idea of the autonomy of science from religious beliefs. On the basis of these two indicators, four profiles were defined as follows:

– Profiles 1 and 2 (non-secularized conceptions of science): at least one of the three means in fideism, classic concordism or reversed concordism is ≥ 2.5 (on a scale of -10 to +10).

– Profile 3 (neither non-secularized nor secularized conception of science): the means in fideism, classic concordism and reversed concordism are all < 2.5 (on a scale of -10 to +10) and the mean in the autonomy of science is also < 2.5 (again on a scale of -10 to +10).

---

4. Seven European countries (Belgium, France, Germany, Italy, Greece, Poland, Spain), the three Maghreb countries, three sub-Saharan African countries (Côte d'Ivoire, Democratic Republic of Congo, Senegal), two Latin American countries (Argentina and Peru) and Turkey. Within each country, schools were selected according to three criteria – urban/rural, public/denominational, science/non-science streams – in order to ensure a certain diversity of cases, if not strict representativity. In all countries, the questionnaire was administered with the double anonymity of both the students and the schools.

5. For better readability, the results obtained on the Likert scales on a scale of 1–6 were recoded as follows: (1) = -10; (2) = -6; (3) = -2; (4) = 2; (5) = 6; (6) = 10. This was done in order to make the degrees of disagreement (negative values) or agreement (positive values) more apparent, on a scale ranging from -10 to +10.

6. In accordance with the reference model, which distinguishes between a secularized conception in the strict sense and an excluding form of secularism, a third indicator should also be retained: the absence of scientistic conceptions. However, since the previous version of the model included only six positions instead of seven, the questionnaire did not include items relating to scientism. This position could therefore not be taken into account here.

– Profile 4 (secularized conception of science): the means in fideism, classic concordism and reversed concordism are all <2.5 and the mean in the autonomy of science is ≥2.5.

Table 3.2 summarizes these different cases and also explains the difference between profiles 1 and 2.

Fideist and concordist conceptions	Low autonomy (< 2.5)	Average autonomy (2.5 - 4.9)	High autonomy (≥ 5)
One or more of the three averages is or are ≥ 5	Profile 1	Profile 1	Profile 2
One or more of the three averages is or are ≥ 2.5 and < 5	Profile 1	Profile 2	Profile 2
All three averages are < 5	Profile 3	Profile 4	Profile 4

Table 3.2. *Non-secularized/secularized conceptions of science: definition of student profiles*

Profile 1 is qualified as "non-secularized in the strict sense" insofar as the student adheres to fideist and/or concordist conceptions, but adheres to the idea of the autonomy of science either little or not at all. Profile 2 is qualified as "non-secularized hybrid form" because the student adheres to both fideist and/or concordist conceptions, as well as to the idea of the autonomy of science. This may seem contradictory, but the individual may, for example, be fideist and/or concordist in some areas (where they perceive similarities with their own religious beliefs) and in favor of the autonomy of science in others. Profile 3, described as "neither secularized nor non-secularized", may correspond to students who are either ambivalent and hesitate between the two positions (non-secularized/secularized), or show themselves to be indifferent, or even reject them both, for reasons that remain to be clarified. Profile 4, finally, is qualified as "secularized" as the student does not adhere to fideist and/or concordist conceptions and is in favor of the idea of the autonomy of science with respect to religious beliefs.

In addition to questions about students' views on the secularization of science, the questionnaire also included one question about their beliefs about religion, another about the status accorded to Scripture and another about the status accorded to the character of Adam mentioned in the Bible and the Qur'an.

Table 3.3 presents students' conceptions[7] regarding the secularization of science[8]. Based on their country and religious beliefs, 31 groups of students could be defined[9]. On the basis of the most frequent profile(s) within these groups, defined with reference to Table 3.2 (conceptions regarding the secularization of science), they were grouped into broader categories (blocks).

In block 1, profile 1 (non-secularized conceptions in the strict sense) is the most frequent and is also statistically more frequent than profile 2. In block 2, the non-secularized profiles (strict form and hybrid form) are the most frequent, without any statistically significant difference between them. In block 3, profiles 1–3 are the most frequent. In block 4, profile 4 (secularized) is the most frequent, but does not form the majority, and finally, in block 5, profile 4 is the majority.

For comparison, the last column of the table shows the percentage of students in each group who think that Adam was "a real character". (The other two options were "the mythical hero of a symbolic story" and "I don't know".)

Most frequent profiles			Student profiles in secularizing science				Comparison of profiles 1 and 2		Adam PR
	Conviction	(n)	Profile 1	Profile 2	Profile 3	Profile 4	Chi^2	S.	
Block 1 (profile 1)									
Algeria	M	(200)	**85.5**	13	/	1.5	106.7	.000	97.5
Morocco	M	(297)	**69**	23.2	4.7	3	67.5	.000	94.9
Peru	P	(47)	**63.8**	21.3	10.6	4.3	10	.002	70.2
Belgium	M	(248)	**59.7**	23.8	6	10.5	36.3	.000	82.7
Congo	P	(76)	**57.6**	34.2	1.3	6.6	4.6	.03	85.5
Turkey	M	(138)	**57.2**	21	18.1	3.6	23.1	.000	83.3
Côte d'Ivoire	P	(157)	**56.7**	33.1	6.4	3.8	9.7	.002	81.5

7. Within the Belgian sample, in order to obtain a significant number of students of different beliefs, we went to schools with different profiles, depending on the belief(s) represented (e.g. schools with a high proportion of Muslim or Protestant students, etc.). It should also be noted that the Protestant students in Belgium are mostly from Central African countries.

8. This table only shows the results for students in general education streams, as technical or vocational streams do not exist in all these countries for this age group.

9. Not included in this table are so-called "theist" students, who say they "believe in a higher principle, but do not adhere to any religion", religions that are very weakly represented within each country (e.g. Judaism, Buddhism, Hinduism), or agnostic or atheist students from countries where they are present, but in very small numbers (Peru, Poland, Tunisia).

Côte d'Ivoire	M	(91)	**52.7**	27.5	15.4	4.4	8	.005	78
Côte d'Ivoire	C	(264)	**49.6**	34.5	8	8	7.2	.01	78
Congo	C	(50)	**48**	28	10	14	2.6	(.10)	92
Argentina	P	(65)	**47.7**	24.6	6.2	21.5	4.8	.03	69.2
Belgium	P	(115)	**45.2**	24.3	13	17.4	7.2	.007	69.6
Poland	C	(308)	**42.2**	25.6	22.7	9.4	12.4	.000	36.5
Block 2 (profiles 1 and 2)									
Senegal	M	(369)	**50.4**	**43.1**	4.1	3			82.7
Senegal	C	(39)	**38.5**	**46.2**	5.1	10.3			64.1
Tunisia	M	(149)	**37.6**	**39.6**	3.4	19.5			74.5
Block 3 (profiles 1–3)									
Peru	C	(263)	**27.4**	**26.2**	**29.7**	16.7			52.1
Block 4 (profile 4, the most frequent, but not the majority)									
Argentina	C	(241)	22.4	27	17	**33.6**			33.6
Greece	0	(197)	20.8	21.8	23,4	**34**			27,4
Block 5 (profile 4)									
France	C	(178)	8.4	18.5	12.4	**60.7**			14
Spain	C	(276)	8	10.5	20.3	**61.2**			14.9
Italy	C	(110)	8.2	20	10	**61.8**			16.4
Germany	C	(55)	3.6	3.6	29.1	**63.6**			16.4
Belgium	C	(188)	9.6	11.2	13.3	**66**			16
Argentina	AA	(124)	5.6	4	18.5	**71.8**			0.8
Germany	P	(106)	3.8	3.8	15,1	**77.4**			15.1
Belgium	AA	(293)	2.7	3.8	10.6	**82.9**			3.4
Spain	AA	(161)	3.1	3.7	7.5	**85.7**			0.6
France	AA	(162)	0.6	4.3	8	**87**			2.5
Italy	AA	(77)	1.3	6.5	3.9	**88.3**			3.9
Germany	AA	(73)	/	2.7	4.1	**93.2**			0

Legend: M = Muslim, P = Protestant, C = Catholic, O = Orthodox, AA = Agnostic or Atheist.

Table 3.3. *Conceptions of students in the senior class (general tracks) regarding the secularization of science*

Although they must be taken with great caution, especially when the sample sizes are particularly small, two important results deserve to be highlighted.

First observation: within these samples, a secularized conception of science clearly appears to be in the minority among the religious students. As a result, among the believing students from these 16 countries, i.e. 25 groups, only six have a predominantly secularized conception of science, in the sense that more than 50% of the students within these groups have a secularized conception of science (profile 4). This is block 5: Catholic students from Belgium, France, Germany, Italy and Spain, as well as Protestants from Germany. All the other groups thus have, to varying degrees (which the different blocks tend to reflect), predominantly non-secularized conceptions of science (profiles 1 and 2) or indeterminate conceptions (profile 3). This shows the extent of the problem. Nearly four centuries after Descartes and Galileo, a secularized conception of science is not at all self-evident for a majority of religious students in these 16 countries. It should be noted that the questions asked in this part of the questionnaire (positions between science and religious beliefs) did not concern the theory of evolution, which shows that the problem has a much more general scope.

Second observation: with one notable exception (the case of the Catholic students from Poland in this sample), there appears to be a very clear link between the belief in the reality of the character of Adam, testifying to a literal or at least concordist reading of the Scriptures, and a non-secularized conception of science. If we exclude the case of Poland, the proportion of students believing that Adam is a character who really existed thus varies between 69% and 97% in blocks 1 and 2. On the contrary, the majority of believing students in block 5 attribute a symbolic rather than a literal status to the Scriptures, in the sense that for them, for example, the story about Adam is aimed at giving meaning to the existence of human beings, but not at explaining his origin factually and historically.

With regard to the particular case of Poland, a possible key to understanding these paradoxical results could be the distinction, in the sociology of religions, between "belonging" and "believing" (Davie 1997). A portion of these students could have a rather strong sense of belonging, for identity-related and cultural reasons, to the Catholic Church and its conceptions, generally considered to be rather conservative in Poland, independently of their own relationship to the Scriptures and/or their personal beliefs[10]. This point would, of course, merit further study.

10. According to S. Urbanski (personal communication): "In Poland, for the last 30 years, secularism has been presented by a large part of the clergy and political staff as an emanation of communism (...). It might be possible to understand the rejection of a secularized

3.4. Some factors that may influence a secularized view of science and students' conceptions of evolutionary theory

The results of the international survey, cited above, clearly showed a negative relationship between adopting a secularized view of science and believing that Adam was a real historical figure rather than a mythological literary figure. In this section, we will focus on the effect of two variables regarding students' relationship to Scripture and their course of study on:

– on the one hand, their conceptions concerning the secularization of science;

– on the other hand, some indicators of their conceptions about the theory of evolution.

It is indeed reasonable to think that a literal reading of the Scriptures and a non-scientific course of study are unfavorable factors a priori for a secularized conception of science, as well as for an appropriation of the theory of evolution. On the contrary, a more distanced reading of the Scriptures (considering them symbolic narratives) and a more advanced scientific education should be a priori favorable factors. We will therefore examine the presumed effect of these two factors with four groups of students in the final year of secondary school (Muslims, Protestants, Catholics, agnostics–atheists) in French-speaking Belgium. This country was chosen because we have a relatively large sample size in the French-speaking part of the country, including four types of belief and three courses of study, i.e. 12 possible cases.

With regard to the relationship to Scripture, in addition to the question about the status of the character of Adam, the following items were proposed:

1) "To me, the sacred Scriptures are a narrative of divine origin, actually, historically, materially explaining the origin of man. This explanation must be accepted as is."

2) "For me, the sacred Scriptures are a narrative of divine origin, really, historically, materially explaining the origin of man. This narrative, formulated according to the language of a time is to be completed or interpreted, based on new knowledge."

3) "For me, the sacred scriptures are a narrative of divine origin, symbolic, aimed not at materially explaining the origin of man, but at giving it meaning."

---

conception of science in the context of this history as very much linked to Polish identity, where the very idea of separation of the 'magistracies' is frowned upon and brings back bad memories."

4) "To me, sacred scriptures are myths produced by men to answer some of their existential questions."

5) I choose both answers 3 and 4.

6) I don't know.

Answer 1 was coded: "type 1 realistic narrative" (R1) corresponding to a literal type of reading. Answer 2 was coded "type 2 realistic narrative" (R2), testifying to an overall concordist conception, characterized by a first degree of distancing from a strictly literal reading. Answers 3–5 were coded: "symbolic and/or mythical type narrative" (S/M). Finally, answer 6 will be noted as "?".

Regarding the study path, three paths have been selected:

– technical and vocational education with a scientific training of less than 6 hours per week, referred to as "TV"[11];

– "basic sciences" general education (3 hours per week of physics, chemistry, biology), referred to as "GBS";

– "strong sciences" general education (6 hours per week of physics, chemistry, biology), referred to as "GSS".

The views on the secularization of science are the four profiles presented in Table 3.4. With respect to evolutionary theory, given the length of time it took to administer the questionnaire on positions towards science and religious beliefs, this part of the survey had to be relatively short. It was limited to 10 questions related to the following four indicators:

– indicator 1: the rejection of a fixist conception (and the adoption of an evolutionary conception), for animals, on the one hand (items 54 reversed, item 56), and for humans, on the other hand (item 62);

– indicator 2: recognition of the existence of common ancestral species, both among animals (item 55, item 57 reversed) and also common to humans and animals (item 61 reversed, item 64);

– indicator 3: the rejection of a finalist view of evolution and the rejection of a Lamarckian-type explanatory mechanism (item 58 reversed);

11. There is also technical education with 6 or more hours of science, but this is very poorly represented in this sample and will not be considered.

– indicator 4: knowledge of two key elements of the Darwinian explanatory mechanism: recognition of the role of chance and natural selection, for animals (item 59) and for humans (item 66).

The instructions invited the students to distinguish, each time, between what they "know" about the theory of evolution and what they "believe", by inviting them, for each question, to answer in two distinct ways: on the one hand, with reference to the theory of evolution developed by Darwin and, on the other hand, according to their personal beliefs. The wording of this question is debatable, because in all rigor, there is no reason to believe or not in the theory of evolution. It is not a matter of belief, but of scientific knowledge. However, in the vocabulary of students, it is common to oppose "knowledge" of the theory of evolution and "personal beliefs" and to declare, for example, to "know but not believe". Our goal with this question is to examine the extent to which students accept or reject elements of evolutionary theory on the basis of "personal beliefs".

It should be noted that the term "personal beliefs" is to be understood in a broad sense and not limited to religious beliefs. An atheist student may also have "beliefs", in the sense of personal conceptions, to which they give some plausibility or credibility, but which they are unable to prove or demonstrate, and are not sure of their accuracy. Learning to distinguish between "knowledge" and "belief" is an important educational issue. In particular, in situations where we do not have complete control over the information, it is essential to be aware of the limits of our knowledge and, if necessary, to "know that one believes" rather than "believe that one knows" (Gagnon 2010, p. 87).

The questions asked were:

Please answer the following questions:

– (A) with reference to the theory of evolution developed by Darwin, checking each time "Y" (yes), "N" (no) or "?" (I don't know);

– (B) according to your personal beliefs, checking each time "Y" (yes), "N" (no) or "?" (I don't know).

The items for which the "correct" answer is "no" are 54A, 57A, 58A, 61A and 62A. Some of the questions (e.g. 57 and 62) may seem surprising, if not completely out of place, if this were an academic assessment test. The purpose here is to explore some of the students' conceptions.

	(A) According to Darwinian evolutionary theory			(B) According to your personal beliefs		
	Y	N	?	Y	N	?
Item 54: "Animal species have not changed from their origin to the present."						
Item 55: "Many animal species, now different, are descended from common ancestral species."						
Item 56: "Animals are the result of a long evolution of living things, starting with the first of them, the unicellular organisms."						
Item 57: "The animal species were created by God separately, independently of each other."						
Item 58: "Animal species have evolved over time to adapt to their environment. Genetic mutations have occurred in response to environmental pressures."						
Item 59: "Animal species have evolved over time as a result of random genetic mutations (i.e., unrelated to environmental pressures). Individuals carrying more favorable mutations, in a given environment, have overall had more offspring."						
Item 61: "Man appeared on Earth separately, independently of the different animal species."						
Item 62: "God created man as he is today, as it is written in the sacred texts."						
Item 64: "Present-day humans and chimpanzees are descended from common ancestral species."						
Item 66: "The emergence of the human species was as unlikely as the emergence of any other species."						

Table 3.4. *Questions asked about the theory of evolution*

We will first examine the effect of two factors likely to influence students' conceptions of the secularization of science: the course of study and the relationship to sacred Scripture. We will then examine the effect of three factors likely to influence students' conceptions of the theory of evolution, distinguishing each time their degree of "knowledge" of it, based on the questions asked, from their degree of "agreement" with reference to their personal beliefs. These three factors are: the course of study; their conceptions of the secularization of science; and their relationship to Scripture.

3.4.1. *Factors likely to influence a secularized conception of science*

3.4.1.1. *The course of study*

As Table 3.5 shows, the proportion of students with a secularized conception of science (profile 4) is very low, whatever the stream of study, among Muslim students (between 8 and 15%) and Protestants (between 11 and 26%). The differences observed between the two streams are, in both cases, statistically insignificant (using the Chi² frequency comparison test) and we can therefore conclude that the stream of study has no effect.

On the other hand, among Catholic students, the probability of having a secularized conception of science increases from 36% in technical and vocational education (TV) to 63% in general basic science education (GBS) and to 74% in general strong science (GSS). Among agnostic-atheist students, these proportions are 61%, 80% and 91%. The effect of the course of study is statistically significant here and the effect size (Cramer's V) can be described as "medium", referring to Cohen's (1988) markers[12].

	Muslims			Protestants			Catholics			Agnostics–atheists		
	(n)	1-3	4	(n)	1-3	4	(n)	1-3	4	(n)	1-3	4
TV	(133)	91.7	8.3	(66)	89.4	10.6	(156)	64.1	35.9	(147)	39.5	60.5
GBS	(92)	84.8	15.2	(63)	82.5	17.5	(114)	36.8	63.2	(166)	20.5	79.5
GSS	(93)	88.2	12	(38)	73.7	26.3	(74)	25.7	74.3	(127)	9.4	90.6
Chi² (2df)	2.6 (N.S)			4.3 (N.S.)			35.6 S.000			32.5 S.000		
V Cramer	0.09 (N.S.)			0.16 (N.S.)			0.33 S.000			0.28 S.000		

Table 3.5. *Percentage of students with a secularized view of science (profile 4), by major and religious beliefs*

We can also observe, among Catholic or agnostic-atheist students, that the differences are more marked between the TV stream and the two general streams than between the GBS and GSS streams, which suggests that this is not only an effect of scientific training, but more broadly of "general culture", which is more present in the academic disciplines of the general streams than in TV. However, it is also necessary to take into account another possible effect: that of differences in recruitment between streams. As Table 3.6 shows, the proportion of students with at least one parent having completed higher education tends to be lower in TV than in the general streams within the four groups, illustrating the well-known phenomenon

12. *Cohen*'s *(1988)* tags*: 0.10 – small size/0.30 – medium size/0.50 – large size.*

of the reproduction of social inequalities by the school. This proportion also tends to be lower, when all streams are taken together, among Muslim students and, to a lesser extent, among Protestants, compared to Catholic or agnostic-atheist students. All of these factors should be kept in mind when interpreting the results.

Fields of study	Muslims		Protestants		Catholics		Agnostics–atheists	
	(n)	And. parents	(n)	And. parents	(n)	And. parents	(n)	And. parents
TV	133	35.6	66	54.2	156	52.7	147	62.5
GBS	92	62.6	63	63.3	114	82.3	166	82.6
GSS	93	56.7	38	83.8	74	86.1	127	83.2
	318		167		344		440	

Table 3.6. *Percentage of students with at least one parent who has been in higher education, by field of study and personal religious beliefs*

3.4.1.2. *The relationship to the sacred Scriptures*

Table 3.7 shows the percentage of students with a secularized conception of science (profile 4) according to their relationship to the Scriptures. It appears – in all four groups (Muslims, Protestants, Catholics, agnostics–atheists) – that students who attribute a symbolic and/or mythical status to the Scriptures are proportionally more likely to have a secularized conception of science. The effect size (ranging from 0.27 to 0.42) can be described as "medium" in reference to Cohen's (1988) markers.

	Muslims			Protestants			Catholics			Agnostics–atheists		
	(n)	1–3	4	(n)	1–3	4	(n)	1–3	4	(n)	1–3	4
R1	(122)	96.7	3.3	(55)	96.4	3.6	(29)	75.9	24.1			
R2	(96)	89.6	10.4	(42)	88.1	11.9	(38)	65.8	34.2	(10)	80	20
?	(92)	87.5	12.5	(14)	64.3	35.7	(46)	56.5	43.5	(57)	61.4	38.6
S/M	(56)	69.6	30.4	(47)	70.2	29.8	(228)	37.7	62.3	(362)	16	84
Chi^2 (3df)	27.9 S.000			17.1 S.001			24.6 S.000			74.5 S.000		
V Cramer	0.30 S.000			0.33 S.001			0.27 S.000			0.42 S.000		

Table 3.7. *Proportion of students with a secularized view of science (profile 4) by their relationship to sacred Scripture and their beliefs about religion*

The observation that students who attribute a symbolic and/or mythical status to the Scriptures are more likely to have a secularized conception of science is not a great surprise, but it nevertheless has the merit of showing the importance of this variable. In comparison, the effect of the field of study is more limited, affecting only Catholic and agnostic-atheist students.

3.4.2. *Factors that may influence students' representations (knowledge vs. beliefs) about evolutionary theory*

3.4.2.1. *The course of study*

Table 3.8 shows, on a scale of 0–10 (since there are 10 questions), the degree of "knowledge" and the degree of "agreement" (in reference to their personal beliefs) of the students according to their fields of study and their personal convictions regarding religion. As regards "knowledge" of the theory of evolution, it appears – quite logically and in the four groups (Muslims, Protestants, Catholics, agnostics–atheists) – that the lowest results are those of the TV stream and the highest are those of the GSS stream (the GBS stream obtaining intermediate results). The differences are statistically significant, and the effects can be described as "medium" (between 0.25 and 0.47) based on Cohen's f index[13].

On the other hand, the degree of "agreement" (in reference to their personal beliefs) is very low among Muslim and Protestant students (between 2.4 and 3.3 on a scale of 0–10), and the course of study is therefore not observed as having any effect. This is not the case, on the other hand, for Catholic or agnostic-atheist students, for whom an effect of the course of study can be demonstrated, along the same lines as that observed for the degree of "knowledge". (The effect here is also of medium size.)

In order to analyze the relations between the degree of "knowledge" and the degree of "agreement" in more detail, according to the field of study and the type of belief, a correlation coefficient (Spearman's Rho) was calculated each time. It appears that the correlations between degrees of "knowledge" and "agreement" are strongest among agnostic-atheist and Catholic students, particularly in the general fields of study (between 0.47 and 0.63). However, this is not the case for Muslim or Protestant students in the general streams, for whom no correlation is statistically significant. In other words, an increase in the degree of knowledge is not associated, in their case, with an increase in the degree of agreement.

13. It is interpreted according to the same guidelines as those mentioned above.

Pathways	Muslims			Protestants			Catholics			Agnostics–atheists		
	(n)	C	A	(n)	C	A	(n)	C	A	(n)	C	A
TV	(126)	3.8	2.8	(61)	4.5	3.2	(149)	4.4	4.6	(141)	4.6	6.3
GBS	(83)	4.8	3.2	(61)	5	2.4	(105)	5.4	5.6	(161)	6.5	7
GSS	(88)	5.7	3.3	(33)	6.2	3	(72)	6.4	6.5	(126)	7.4	7.5
AV/Welch		16	1.8		5.3	2,1		15.5	17.4		43.1	11.8
S.		0.000	/		0.001	/		0.000	0.000		0.000	0.000
f Cohen		0.31	/		0.25	/		0.29	0.33		0.47	0.25
rho C/A												
TV	0.23 S.01			0.31 S.02			0.34 S.000			0.54 S.000		
GBS	0.08 NS			0.06 NS			0.47 S.000			0.63 S.000		
GSS	0.18 NS			0.27 NS			0.58 S.000			0.62 S.000		

NB: The statistical test of comparison of means used is the analysis of variance, possibly in the form of Welch's coefficient, when the condition of homogeneity of variances is not met.

Table 3.8. *Levels of "knowledge" and "agreement" with the Darwinian theory of evolution, by field of study and personal religious beliefs*

This table therefore also allows us to highlight interesting discrepancies between the degree of "knowledge" of the students and their degree of "agreement", in reference to their personal beliefs. Among Muslim and Protestant students, the degree of agreement is systematically lower than the degree of knowledge, whatever the field of study. It should be noted that among agnostic-atheist students, it is rather the opposite, as far as the TV stream is concerned. These students would tend to agree with different conceptions of the theory of evolution, even if they do not know the theory precisely. Finally, the closest parallelism between the degree of knowledge and the degree of agreement seems to be among Catholic students. Among the latter, the degree of acceptance of the theory of evolution clearly tends to increase with the degree of knowledge.

It would be interesting, but not feasible in the space allotted to this chapter, to examine how the degrees of "knowledge" and "agreement" evolve for each of the 10 questions asked for each stream and for each of the four groups. We will therefore limit ourselves in Table 3.9 to presenting the results for the GSS stream. This is the stream with the highest level of knowledge. Table 3.9 shows, for each question, the percentage of students in the GSS (general strong science) stream who answered correctly. It appears that the questions under indicators 1 (rejection of fixism) and 2 (common ancestral species) are generally well answered by students in this stream in all four groups (Muslims, Protestants, Catholics, agnostics–atheists). However, the situation is quite different for the explanatory mechanisms of evolution. While as regards the items explicitly referred to the theory developed by Darwin, the students ticked "yes" to item 58, expressing a Lamarckian conception of evolution, which is reflected in the very low success rates for this item.

As can be seen, this ignorance of the Darwinian conception of evolution and the confusion with a Lamarckian conception in a finalist view is clearly independent of the religious factor, since it is present in all four groups, including the agnostic-atheists. This finding is important because it shows that the importance of the religious factor should be relativized. Many other factors can constitute a source of important difficulties in the understanding of the theory of evolution.

		Rejection of fixism			Common ancestral species				Explanatory mechanisms		
	(n)	**Q.54**	Q.56	**Q.62**	Q.55	**Q.57**	**Q.61**	Q.64	**Q.58**	Q.59	Q.66
M		69	78	61	87	58	58	81	7	54	33
P		69	81	81	79	61	70	88	0	42	64
C		78	81	80	83	75	75	80	10	43	42
AA		90	94	92	91	84	83	85	14	51	52

NB: Items in bold are those where the expected answer is "no".

Table 3.9. *Knowledge: percentage of correct responses of GSS students to different questions*

If we now look at the degree of "agreement" (Table 3.10), we can see that the percentages of Catholic and agnostic-atheist students are very close in "knowledge" and "agreement" for all of the questions under the first two indicators. Among Muslim and Protestant students, on the other hand, these data confirm a form of dissonance between their knowledge and their personal beliefs: their degree of agreement tends to be lower overall than their degree of knowledge. The discrepancies are particularly pronounced among these two groups for certain particularly sensitive questions: question 62 (rejection of a fixist conception for the human being), questions 61 and 64 (ancestral species common to humans and other species); and to a lesser extent for other questions relating to animals (questions 56 and 57). These results tend to suggest, however, that these students would be able to distinguish between knowledge of Darwinian theory itself and their degree of agreement with it, with reference to their personal beliefs.

Regarding question 58, a majority of students (between 62% and 82% depending on the group) agreed with this proposition, even though it was incorrect, which explains the low rates observed for this question in the four groups.

		Rejection of fixism			Common ancestral species				Explanatory mechanisms		
	(n)	**Q.54**	Q.56	**Q.62**	Q.55	**Q.57**	**Q.61**	Q.64	**Q.58**	Q.59	Q.66
M		55	55	21	58	28	23	23	22	38	21
P		57	35	23	60	15	32	27	31	44	32
C		79	83	80	84	74	68	78	14	48	52
AA		86	92	94	91	91	91	84	15	51	59

Table 3.10. *Agreement: percentage of GSS students' responses agreeing with the answers deemed correct*

3.4.3. *The effect of views on the secularization of science on the degree of knowledge and agreement with the theory of evolution*

Table 3.11 presents the degree of "knowledge" and "agreement" of students regarding the Darwinian theory of evolution, according to their relationship to the secularization of science. With the exception of the special case of Protestant students in "knowledge", it appears that students with a secularized view of science (profile 4) tend to have both higher levels of knowledge of and agreement with Darwinian evolutionary theory than other students. (These effects tend to be medium-sized, again with reference to Cohen's tags.)

Profiles	Muslims			Protestants			Catholics			Agnostics–atheists		
	(n)	C	A	(n)	C	A	(n)	C	A	(n)	C	A
1	(161)	4.7	2.6	(77)	4.7	2.3	(38)	4.5	3.8	(11)	5.2	5.7
2	(53)	4.6	3.1	(31)	5.3	2.8	(35)	4.5	5.1	(11)	4.7	6.1
3	(47)	3.6	3.5	(20)	5	2.8	(75)	4.1	4.7	(76)	4.7	6
4	(36)	5.7	4.1	(27)	5.7	4.5	(178)	5.8	6.1	(330)	6.5	7.2
AV/ Welch		4.7	5.1		1.1	5.3		8.2	12.4		8.5	7.8
S.		0.004	0.003		/	0.003		0.000	0.000		0.000	0.001
f Cohen		0.20	0.25		/	0.40		0.27	0.35		0.29	0.29

Table 3.11. *Means in "knowledge" and "belief" about the theory of evolution according to conceptions about the secularization of science*

3.4.4. *Effect of relationship with the Scriptures on the degree of knowledge of and agreement with the Darwinian theory of evolution*

Table 3.12 presents students' levels of "knowledge" and "agreement" in terms of Darwinian evolutionary theory as a function of their relationship to Scripture. It appears that students who attribute symbolic and/or mythical status to Scripture have both higher levels of "knowledge" and "agreement" in all four groups. (The effect size is medium to large, again with reference to Cohen's markers.)

	Muslims			Protestants			Catholics			Agnostics–atheists		
	(n)	C	A	(n)	C	A	(n)	C	A	(n)	C	A
R1	(112)	4.1	2.6	(51)	4.2	2	(27)	3.4	3.6			
R2	(88)	5.1	3.1	(39)	5.2	2.4	(36)	4.3	4	(10)	4.4	4.3
?	(31)	3.1	3	(13)	5.1	3.9	(43)	4.3	4.4	(55)	4	5.9
S/M	(55)	5.8	4	(43)	6	4.3	(218)	5.6	6	(355)	6.5	7.2
AV/Welch		10.7	4.3		4.3	12.9		7.9	17.9		17.6	11.3
S.		0.000	0.01		0.01	0.000		0.000	0.000		0.000	0.000
f Cohen		0.3	0.25		0.29	0.55		0.27	0.42		0.35	0.29

Table 3.12. *Means in "knowledge" and "belief" in terms of the theory of evolution according to the relationship to the Scriptures*

These results clearly show that attributing a symbolic and/or mythical status to the sacred texts and having a secularized conception of science are two associated variables, both having a positive and significant effect on both the degree of knowledge and the degree of agreement with the theory of evolution. This is true for all four groups (Muslims, Protestants, Catholics, agnostics–atheists), showing the importance of these two variables.

3.5. Some conclusions and avenues for reflection at the pedagogical and didactic level

Even if the results must be considered with caution, especially when the sample size is small, two findings of the international survey are worth highlighting. First observation: in this first quarter of the 21st century, among the religious students of these 16 countries, i.e. 25 groups, only six have a predominantly secularized conception of science, in the sense that more than 50% of the students in each of these six groups have a secularized conception of science (profile 4): the Catholic students of five European countries (Germany, Belgium, Spain, France, Italy) and the Protestant students of Germany.

Without the results of similar surveys carried out at other times, it is difficult to estimate what the influence of the rise of fundamentalist movements within different religions and in different parts of the world since the 1980s, for example, has been. The second important observation is that, whatever the religion (Catholicism, Islam, Protestantism), belief in the historical reality of the figure of Adam tends to be associated with a non-secularized conception of science. This shows the importance of the relationship to the Scriptures in the adoption or non-adoption of a secularized conception of science.

To investigate these results further and to examine the implications for learning about evolutionary theory, a second survey was conducted, this time in French-speaking Belgium, with four groups of students (agnostic-atheists, Catholics, Muslims, Protestants) and within three streams of study. Ten questions were added on the degree of the students' knowledge of the theory of evolution and on their degree of agreement (taking into account their personal beliefs) with it. Three findings are worth noting from this second survey.

The first observation is that the religious factor is far from sufficient to explain all of the difficulties in understanding the theory of evolution encountered by students. As other similar surveys have already shown for France (Fortin 2009; Peterfalvi 2009; Guillot 2015), students in French-speaking Belgium, including in the general "strong sciences" stream, tend to have a Lamarckian conception of evolution and to confuse it with the Darwinian conception (which was the subject of the question). This confusion is clearly independent of the religious factor as it is present in all four groups (Muslims, Protestants, Catholics, agnostics–atheists). A first possible explanation is the difficulty for students in distinguishing an "evolutionary mechanism" as a non-finalized cause (in reference to Darwin's theory) from a "physiological mechanism" with predictable effects (Fortin 2009). In addition, there are ambiguities in the school curricula concerning the notion of "adaptation", with formulations that, although avoiding any finalism, are close to a technological discourse, with expressions such as "is facilitated by", "allows" and "results" (Orange 2009). A second possible explanation concerns teaching practices. These are based on a conception that is sometimes very empirical, not clearly presenting the different interpretative models (e.g. Cuvier, Lamarck, Darwin) and their arguments, and therefore not allowing students to make the expected distinctions (Fortin 2009). Cobut (2009) also points out that, in some Belgian programs, the distinction between Lamarck's and Darwin's theories is only presented as a "hook", when it is actually quite fundamental.

Second, there were significant discrepancies among some students between their degree of knowledge of the theory of evolution and their degree of agreement with it, given their personal beliefs: Muslim and Protestant students (all streams) tended to have a degree of agreement that was lower than their degree of knowledge, whereas agnostic-atheist students, albeit only in the technical–vocational stream, tended to have a degree of agreement that was higher than their degree of knowledge. These discrepancies could be explained, at least in part, as Fortin (2009) and Peterfalvi (2009) suggest, by an ahistorical conception of time and fixist ideas (reinforced most often by certain religious convictions) in the first case, or, on the contrary, by its too easy acceptance, assimilating evolution to a form of general metamorphosis of the living world, without taking into account the mechanisms of rupture or speciation, in the second case. They could also be explained by processes of identity (self-)assignment (Mathieu 2011), leading some students to accept or, on the contrary, to reject – on principle – the theory of evolution.

The third important finding: attributing a symbolic and/or mythical status to the Scriptures and having a secularized conception of science are two associated variables, both having a positive and significant effect on the degree of knowledge of the theory of evolution and on the degree of agreement with it. This is true for all four groups (Muslims, Protestants, Catholics, agnostics–atheists).

On the basis of these observations, several lines of thought, from a pedagogical and didactic point of view, seem to us to be possible. The first, which we will not go into here, concerns the importance, in learning the scientific approach, of confronting observational data with interpretative models and therefore of presenting these models explicitly to the students, which a purely empirical conception tends to neglect.

A second avenue relates to the distinction between "knowledge" and "belief". Learning to distinguish between "knowledge" and "belief" is an important educational issue. It is indeed essential to be aware of the limits of our knowledge and, if necessary, to "know that one believes" rather than "believe that one knows" (Gagnon 2010, p. 87). However, it is important to distinguish between a belief that is simply the result of a lack of information, precision or knowledge about a given topic, and a religious type of belief. Religious belief is a particular case of belief that intends to refer to a transcendent foundation that reason alone is not sufficient to admit. It requires an act of faith. By definition, without this act of faith, there is no religious belief either. Consequently, while it may seem relevant, at first, for students to be able to express possible discrepancies between their degree of knowledge of the theory of evolution and their degree of agreement with it, on the basis of their personal beliefs, it is nevertheless highly desirable to gradually bring students to the idea that there is no reason to oppose scientific knowledge and religious beliefs, nor to seek to validate one by the other, insofar as the two belong to fundamentally different discursive registers. In this sense, there is no need to believe or not to believe in the theory of evolution. This leads to a reflection on the criteria of demarcation between science and religious beliefs, and consequently also to a reflection on the secularized/non-secularized character of science, as well as on the status of the Scriptures.

Regarding this first component (demarcation criteria and secularized/non-secularized conceptions of science), students should be given informed knowledge of the purpose and limits of science (Aroua et al. 2012). Some elements of the history of science are therefore essential: explaining to students how science has been constructed historically, as well as the methodological choices made, especially since the advent of "modernity"; the limitation of the ambitions of science in the search for "efficient causes" and not for "ultimate causes", as proposed by Descartes in the 17th century; the distinction between "knowledge", "belief" and "opinion" proposed by Kant in the 18th century; or the principle of "refutability" enunciated by Popper in the 20th century, in spite of its limits and the difficulties in interpretation to which it can give rise. We can possibly evoke as a precursor the "principle of parsimony" in the explanatory approach (stated by D'Occam in the 14th century). The understanding of these methodological choices and their rationale should lead to the idea of incommensurability between the scientific and religious registers and to that of autonomy between the scientific explanatory register and religious beliefs[14]. The act of learning of these principles benefits from it being done, initially, on ideologically insensitive subjects, so that they will be well acquired by the time more sensitive subjects are to be tackled. Moreover, the use of the historical approach can also be an opportunity to emphasize and value, for the students, the contribution of different cultures to the development of science[15].

At the same time, with regard to the second aspect (relationship to the Scriptures), it seems just as necessary, in a register that can be one of teaching religious facts, to make students aware of the diversity of readings and interpretations of the Scriptures depending on the historical and cultural context, so that those who would be tempted by a literal reading can distance themselves from it.

---

14. In contrast, a religious belief is about "ultimate causes" and not about "efficient causes", and it is not required to be parsimonious or refutable.

15. For example, a traveling exhibition organized in French-speaking Belgium in 2007 on the theme of "Discovering the Golden Age of Arab Sciences" allowed many classes to become aware of the richness of this contribution and strengthen the bonds of esteem and respect between different cultures.

Finally, let us remind ourselves that attitudes towards science and/or religions do not only depend on strictly rational factors, but that emotions and the relational context also play a part: namely, the school climate, a feeling of being recognized or, on the contrary, discriminated against within the school and society, and so on. Students who feel devalued, or even discriminated against, on the basis of their personal convictions and/or cultural origin could, in return, reject scientific content that they consider threatening to their identity, in the name of an ontological overvaluation of the latter (Heine 2008) or of identity self-assignment processes (Mathieu 2011). This is another reason to value the contribution of different cultures to the development of science.

3.6. References

Alexander, D.R. (2007). Models for relating science and religion. *Faraday Papers*, 3, 1–4.

Aroua, S. (2009). Enseignement de l'évolution et contexte socioculturel, le cas de la Tunisie. In *L'évolution du vivant : un enseignement à risque ?*, Coquidé, M. and Tirard, S. (eds). Vuibert & Adapt-snes, Paris.

Aroua, S., Coquidé, S., Abbes, S. (2012). Controverses dans l'enseignement de l'évolution. Questions de recherche sur les stratégies d'intervention en classe et dans la formation. *Revue de didactique des sciences et des technologies*, 5, 47–76.

Barbour, I.G. (1990). *Religion in an Age of Science.* SCM Press, London.

Billingsley, B., Taber, K., Riga, F., Nedwick, H. (2013). Secondary schools students epistemic insight into the relationships between science and religion – A preliminary enquiry. *Research in Science Education*, 43(4), 1715–1732.

Chabchoub, A. (2001). Rapports aux savoirs scientifiques et culture d'origine. In *Les jeunes et le savoir : perspectives internationales*, Charlot, B. (ed.). Anthropos, Paris.

Chaline, J. and Grimoult, C. (2011). *Les sciences de l'évolution et les religions : enjeux scientifiques, politiques, philosophiques et religieux*. Ellipses, Paris.

Clément, P. (2014). Les conceptions créationnistes d'enseignants varient-elles en fonction de leur religion ? *Education et sociétés*, 33, 113–136.

Cobut, M.G. (2009). *Comprendre l'évolution : 150 ans après Darwin*. De Boeck Supérieur, Collection Action, Brussels.

Cohen, J. (1988). *Statistical Power Analysis for the Behavioral Sciences*, 2nd edition. Lawrence Erlbaum, Hillsdale, NJ.

Davie, G. (1997). Believing without belonging: A framework for religious transmission. *Recherches sociologiques*, 3, 17–37.

El Asri, F. (2009). Discours musulman et sciences modernes : un état de la question. In *Adam et l'évolution : Islam et christianisme confrontés aux sciences*, Maréchal, B. and Dasseto, F. (eds). Academia Bruylant, Louvain-la-Neuve.

Fortin, C. (2009). L'enseignement de l'évolution au lycée, entre description et explication. In *L'évolution du vivant : un enseignement à risque ?*, Coquidé, M. and Tirard, S. (eds). Vuibert &Adapt-snes, Paris.

Gagnon, M. (2010). Regards sur les pratiques critiques manifestées par des élèves du secondaire dans le cadre d'une réflexion menée en îlot interdisciplinaire de rationalité. *McGill Journal of Education/Revue des sciences de l'éducation de McGill*, 45(3), 463–494.

Gould, S.J. (2000). *Et Dieu dit "que Darwin soit !"*. Le Seuil, Paris.

Guillot, D. (2015). Le créationnisme est-il seulement une affaire de foi ? Enquête sur l'origine des espèces en France et au Maroc. *Revue européenne des sciences sociales*, 53(1), 227–263.

Haught J.F. (1995). *Science and Religion: From Conflict to Conversation*. Paulist Press, Mahwah.

Heine, A., Van Der Linden, N., Van Den Abeele, C., Licata, L. (2008). Quand les leçons de l'école ne sont pas celles de la maison. Une approche psychosociale du rapport au savoir des enfants d'immigrés musulmans. *Education comparée*, 1, 61–80.

Lambert, D. (1999). *Sciences et théologie – Les figures d'un dialogue*. Presses universitaires de Namur.

Lambert, Y. (2000). Religion, modernité, ultramodernité : une analyse en termes de "tournant axial". *Archives des sciences sociales des religions*, 109, 87–116.

Le Ru, V. (2010). *La science et Dieu. Entre croire et savoir*. Vuibert/Adapt-SNES, Paris.

Martin-Hanssen, L.M. (2008). First-year college students' conflict with religion and science. *Science & Education*, 17, 317–357.

Mathieu, S. (2011). Ce qu'ils en disent : la perception de la théorie de l'évolution par des élèves de college et lycée. In *Théorie de l'évolution et religions*, Portier, P., Veuille, M., Willaime, J.P. (eds). Riveneuve Editions, Paris.

Minois, G. (1990). *L'Eglise et la science. Histoire d'un malentendu*. Fayard, Paris.

O'Brien, T.L. and Noy, S. (2015). Traditional, modern, and post-secular perspectives on science and religion in the United States. *American Sociological Review*, 80(1), 92–115.

Orange, C. (2009). Les fonctions épistémologiques de l'évolution dans les programmes français de sciences de la vie et de la Terre. In *L'évolution du vivant : un enseignement à risque ?*, Coquidé, M. and Tirard, S. (eds). Vuibert & Adapt-snes, Paris.

Perbal, L., Susanne, C., Slachmuylder, J.L. (2006), Evaluation de l'opinion des étudiants de l'enseignement secondaire et supérieur de Bruxelles vis-à-vis des concepts d'évolution (humaine), *Anthropo*, 12, 1–26.

Peterfalvi, B. (2009). Comprendre la théorie de l'évolution, une course d'obstacles ? In *L'évolution du vivant : un enseignement à risque ?*, Coquidé, M. and Tirard, S. (eds). Vuibert & Adapt-snes, Paris.

Portier, P. (2011). L'Église romaine et le darwinisme. Langage de la vérité et art du compromis dans le catholicisme contemporain. In *Théorie de l'évolution et religions*, Philippe Portier, J.-P.W. and Michel, V. (eds). Riveneuve Editions, Paris.

Rasi, H. (2003). La foi, la raison et le chrétien cultivé. *Dialogue*, 15(3), 5–9.

Stenmark, M. (2004). *How to Relate Science and Religion*. Eedermans, Grand Rapids, MI.

Taber, K.S., Billingsley, B., Riga, F., Nedwick, H. (2011). Secondary students' responses to perceptions of the relationship between science and religion: Stances identified from an interview study. *Science Education*, 95(6), 1000–1025.

Urvoy, D. (2006). *Histoire de la pensée arabe et islamique*. Le Seuil, Paris.

Weber, M. (1919). *Le savant et le politique* (translated by J. Freund). Plon Bibliothèque, Paris.

Wolfs, J.L. (2013). *Sciences, religions et identités culturelles. Quels enjeux pour l'éducation ?* De Boeck, Brussels.

Wolfs, J.L. (2018). La concurrence entre savoirs scientifiques et croyances religieuses à l'école : enquête auprès d'élèves de terminale en Belgique francophone et pistes didactiques. *Revue internationale d'éducation. Sèvres*, 77, 103–112.

Wolfs, J.L., Leys, C., Legrand, S., Karnas, D., Delhaye, C., Bouko, C., Zamboni, S. (2014). Les représentations des élèves à propos des différentes postures intellectuelles possibles entre science et croyances religieuses : mise à l'épreuve de la validité de construit d'un questionnaire y afférant. *Mesure et évaluation en éducation*, 37(2), 101–132.

Wolfs, J.L., Delhaye, C., Altepe, C., Vanhove, R., Hassan, W. (2021). Les postures entre science et croyance religieuse : construction d'un modèle d'analyse et comparaison avec les taxonomies existantes. *Recherche en didactique des sciences et des technologies*, 23, 161–181.

Wolfs, J.L., Delhaye, C., Leys, C., Altepe, C., Dini, K., Gauthier, L., Bertrand, L. (2022). Positions on science and religious beliefs across societies: Development of a research instrument and testing of its validity among high school students. *Public Understanding of Science*, 31(4), 394–409 [Online]. Available at: https://doi.org/10.1177/09636625221075787.

Yasri, P. and Mancy, R. (2014). Understanding student approaches to learning evolution in the context of their perceptions of the relationship between science and religion. *International Journal*, 36(1), 24–45.

Yasri, P., Arthur, S., Smith, M.U., Mancy, R. (2013). Relating science and religion: An ontology of taxonomies and development of a research tool for identifying individual views. *Science and Education*, 22, 2679–2707.

PART 2

Teaching Evolution: Educational and Training Issues

This second part, composed of five chapters, is devoted to the content taught in secondary education and teacher training.

The first two chapters focus on the curricular dimension of teaching evolution. Marco Barroca-Paccard's chapter proposes an analysis of French secondary school curricula in the light of Dobzansky's (1973) quotation that "nothing makes sense in biology unless it is in the light of evolution". The author shows that, despite the reference to evolution in curricula related to the history of biodiversity and population approaches in genetics, the theory of evolution is little or not at all mobilized, in particular concerning physiology and environmental issues. Corinne Fortin's work questions the limits of the *Homo* genus by using the case of the chimpanzee, classified by some biologists in the *Pan* genus and others in the *Homo* genus. She studies its curricular re-problematization in secondary and university education, and shows that the curricula reify the *Homo* genus because they fail to take into account the epistemological pluralism at stake. In a mirror effect, the chapter by Magali Fuchs-Gallezot and Corinne Fortin explores the liveliness of epistemological pluralism in evolutionary biology through the contemporary "synthetic theory versus extensive theory" controversy in evolutionary biology. The authors show that this controversy is part of a socio-scientific practice that would benefit from being developed in teacher training to better understand the nature of science. Another controversy, now resolved, is addressed in the chapter by Patricia Crépin-Obert, who studies the evaluation of the age of the Earth in relation to two famous individuals, Darwin and Kelvin, in the context of a continuing education course for science teachers. The training objective is to go beyond a naive epistemology of the construction of scientific knowledge among teachers to develop a more expert epistemology, thereby helping students to problematize scientific knowledge. At the end of this second part, Fabienne Paulin's chapter brings together the ideas laid out in the previous ones through the conceptual landscapes relating to evolution among biology researchers, teachers and students. Based on a comparative analysis, the author shows that the conceptual landscape of teachers is more focused on the effects than on the causes of evolution. The conceptual landscape also appears to be a tool for developing new skills in the teaching of evolution in training situations.

4

If Nothing Makes Sense without Evolution, What Sense Can Be Made of the Biology Content of the Official French School Curriculum?

Marco BARROCA-PACCARD
LirEDD, Haute École Pédagogique Vaud, Switzerland

4.1. Introduction

Biological evolution is a central, unifying and global theme, and is the integrative framework of modern biology (David and Samad 2011). In this sense, biology education should widely integrate evolutionary theory. This fundamental importance of evolutionary theory was highlighted as early as 1973 by Dobzhansky, who wrote an article in the magazine *The American Biology Teacher* with the title "Nothing in biology makes sense except in the light of evolution". This statement calls for the consideration that, in biology, the theory of evolution allows us to understand and organize observations and experiments within a coherent conceptual framework. We will present the place that evolutionary theory can take in the biology school curriculum and then perform a lexicometric analysis of all biology content in the French school curriculum in order to analyze the place of evolutionary theory, as well as its links with the different contents of the earth and life sciences

The Evolution of Life: Teaching, Learning and Training,
coordinated by Corinne FORTIN and Julie GOBERT.

(ELS) school curriculum[1]. We will also analyze the way in which evolutionary content is presented with a focus on the high school senior curriculum (students aged 17–18).

4.2. The place of evolutionary theory in the teaching of biology

Dobzhansky's article shows that biological data, including biological diversity, comparative anatomy, embryology and paleontology only make sense within the framework of evolutionary theory. The author states that:

> Seen in the light of evolution, biology is, perhaps, intellectually the most satisfying and inspiring science. Without that light it becomes a pile of sundry facts, some of them interesting or curious but making no meaningful picture as a whole (Dobzhansky 1973, p. 129).

Dobzhansky's original work refers to the existence of a common evolutionary history of living beings that is presented as indisputable (evolution as a historical process), but does not concern the mechanisms of evolution:

> Evolution as a process that has always gone on in the history of the earth can be doubted only by those who are ignorant of the evidence or are resistant to evidence, owing to emotional blocks or to plain bigotry. By contrast, the mechanisms that bring evolution about certainly need study and clarification (Dobzhansky 1973, p. 129).

However, this idea that nothing makes sense in biology except in the light of evolution has often been extended to consideration of the adaptation of living beings to the environment, which is considered a mechanism of evolution (Griffiths 2009).

Today, acceptance of evolution and recognition of its central and unifying role is affirmed by most scientists, and also by many scientific societies and major educational organizations, which have also issued statements acknowledging the factual reality of evolution and its power to unify the sciences, especially biology (IAP 2006). However, at the beginning of the 21st century, opposition to the teaching of evolution remains present in society, though it has been more than 45 years since Dobzhansky published "Nothing in biology makes sense except in the light of evolution" (Blancke and Kjærgaard 2016). Even though France is little concerned by the challenges facing the teaching of evolutionary theory in schools

1. In France, biology and geology are studied in a school program called "*sciences de la vie et de la terre*", translated here as earth and life sciences (ELS).

(Deniz and Borgerding 2018), reaffirming this centrality of evolution in the teaching of biology remains relevant (Fortin 2009).

In contrast to the debates that sometimes exist in society, there is a consensus among biology educators that the teaching of evolution should be part of the school curricula. There have been debates, however, about the place that such teaching should take (Wouters 2007; McComas 2016). For some, evolution should be restricted to a field of biology distinct from the rest, notably because much research is conducted in physiology or molecular biology by researchers who have little knowledge of evolutionary biology (Wouters 2007). However, this view does not seem to be a majority one. Indeed, while it is quite possible to *teach* biology without taking into account evolutionary theory, it seems impossible to *understand* biology without evolutionary theory:

> Just as nothing makes sense in the biological sciences without evolution, we cannot consider ourselves biology teachers without it either. Biology education is evolution education (McComas 2016, p. 91).

Another topic of debate concerns the grade level at which to address evolutionary theory. In 2002, Edwards proposed to wait until the end of high school (the college level) to teach evolutionary theory, because students would then be better equipped to tackle more advanced questions concerning molecular genetics, cloning and evolutionary biology (Edwards 2002). This proposal by Edwards was met with strong reactions:

> Although I respect Dr. Edwards's right to express his views, the unintended negative consequences of his proposal may far exceed any positive value of the recommendations. [...] Edwards is shouting "Fire!" in a crowded theater. He may mean well, but an evaluation of his position must include a consideration of his intention, the context, and an understanding of the consequences. The evaluation, in my view, fails on the latter count. He presents little understanding of biology education and the consequences of his position (Bybee 2002, p. 616).

From the point of view of biology teaching, the debates are therefore not about the presence or absence of evolutionary theory, but about how to integrate it into the school curriculum.

In this context, it seems necessary to question the place of evolutionary theory in the French science curriculum. Indeed, France is a country where evolutionary theory is very much embedded in the school curriculum (Fortin 2015; Quessada and Clément 2018). However, the place of the theory in the school curriculum does not tell us what role it plays as an explanatory framework for biology content.

The aim of this study is first to highlight the structuring of biological content within the ELS curriculum and to see how the elements related to the theory of evolution are associated with the content of other elements. This will be done through a lexicometric approach of content analysis of science and technology, science education and the ELS curriculum, from 4th grade (aged 8–9) to the high school senior class, integrating the new high school senior curriculum that was introduced in 2019.

Following the overall lexicometric analysis, in the second section, we will propose an analysis of the parts of the school curriculum dealing with the theory of evolution, and analyze in more detail the common scientific education curriculum of the high school senior year entitled "A history of living things" (*Une histoire du vivant*).

4.3. Methodology

In order to highlight the place of the theory of evolution in the scientific curriculum in France, we have created a corpus of research consisting of all the elements of biology present in the official curriculum. The teaching of the theory of evolution is widely present in the scientific and ELS curricula, and the new advanced curriculum for the high school freshman and senior years (published in the special editions of the BO No. 1 of January 22, 2019 and No. 8 of July 25, 2019) ensure that various evolutionary concepts have a prominent place. The reading of all the scientific and ELS curricula from 4th grade to high school senior year was carried out manually and made it possible to select the parts that include at least some biological content (Table 4.1). Only the parts presenting the teaching content were studied and the introductory texts of the curriculum were not included in the analysis. The total corpus consists of more than 170,000 characters and lends itself well to a lexicometric approach that allows for the determination of lexical worlds based on co-occurrences (Reinert 1983). The analyses were performed with the IRaMuTeQ software, R Interface for Multidimensional Text and Questionnaire Analysis (Ratinaud 2009). The initial text corpus is processed to be converted into a set of forms that can be subjected to hierarchical top-down classification; the resulting classes refer to lexical fields (Reinert 1990). In hierarchical clustering, a single class containing all of the text segments is divided step by step into smaller classes to produce a dendrogram. A certain number of complementary analyses can also allow a better interpretation of these classes; in school curricular, the factorial correspondence analysis (FCA) will allow the words and classes to be projected onto axes or planes (called factorials). This will make it possible to visualize those that can be grouped together and those that are opposed. This methodological approach has recently been used to analyze science school curricula (Barroca-Paccard et al. 2018; Fortin 2018).

Level	Official text	Part	Subparts
4th to 6th grade Science and Technology	BO special edition No. 11 from November 26, 2015	The living world, its diversity and the functions that characterize it	
		The planet Earth. living beings in their environment	
7th to 9th grade ELS	BO special edition No. 11 from November 26, 2015	Living things and their evolution	
		Planet Earth, the environment and human action	
		The human body and health	
10th grade ELS	BO edition No. 1 from January 22, 2019	The Earth, life and the organization of living things	The functional organization of living organisms
			Biodiversity, results and stages of evolution
		Contemporary global issues	Geosciences and landscape dynamics
			Agrosystems and sustainable development
		Human body and health	Procreation and human sexuality
			Microorganisms and health
11th grade scientific education	BO special edition No. 1 from January 22, 2019	1. A long history of the material	1.3. A complex structure: the living cell
		2. The Sun, our source of energy	2.3. A biological conversion of solar energy: photosynthesis
			2.4. The thermal balance of the human body
		4. Sound and music, carriers of information	4.4. Hearing music

Level	Official text	Part	Subparts
11th grade specialized ELS course	BO special edition No. 1 from January 22, 2019	The Earth, life and the organization of living things	Transmission, variation and expression of genetic heritage
		Contemporary global issues	Ecosystems and environmental services
			Humanity and ecosystems: ecosystem services and their management
		Human body and health	Genetic variation and health
			The functioning of the human immune system
High school senior year, scientific education	BO special edition No. 8 from July 25, 2019	Theme 1: science, climate and society	1.1. The Earth's atmosphere and life
		Theme 3: a history of life	3.1. Biodiversity and its evolution
			3.2. Evolution as a grid for reading the world
			3.3. Human evolution
High school senior year, specialized ELS course	BO special edition No. 8 from July 25, 2019	The Earth, life and the organization of living things	Genetics and evolution
		Contemporary global issues	From the wild plant to the domesticated plant
		Human body and health	Behavior, movement and the nervous system
			Produce movement: muscle contraction and energy supply
			Behaviors and stress: towards an integrated vision of the organism

Table 4.1. *Elements of French science and the ELS school curricula that address biological concepts*

4.4. Analysis of the entire biology school curriculum

A hierarchical top-down classification was performed on text segments in a simple way and allowed us to build five classes, with associated forms that enable us to interpret these classes (Figure 4.1). This allows us to propose that the French biology school curriculum is organized according to five lexical worlds corresponding to the different themes addressed:

– class 1: ecological and environmental themes (ecosystems, erosion, environment, etc.);

– class 2: evolutionary and genetic themes (population, genetics, evolution, selection, etc.);

– class 3: plant and agriculture-related themes (plants, vegetables, seeds, etc.);

– class 4: immune and health topics (health, pathogens, immunity, vaccines, etc.);

– class 5: human body and physiology themes (nervous, muscular, glucose, etc.).

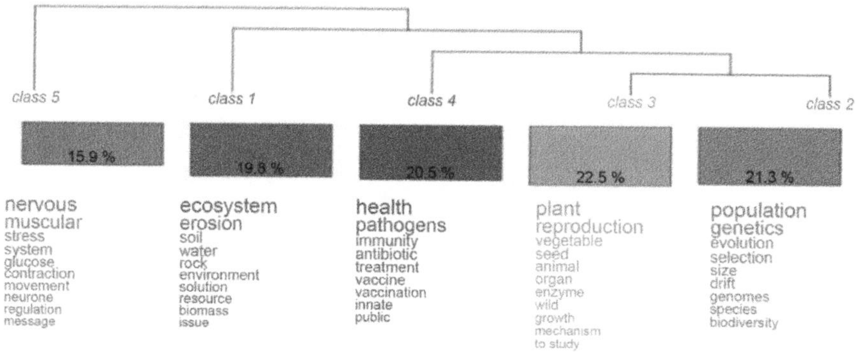

Figure 4.1. *Hierarchical top-down classification using Reinert's method performed on the biology curriculum set (IRaMuTeQ). For a color version of this figure, see www.iste.co.uk/fortin/evolution.zip*

This classification also allows us to show that each of these themes occupies a variable share, which makes it possible to see their relative importance. The evolutionary and genetic themes alone represent 21.3% of the classified text segments and thus constitute one of the important axes of biology teaching in France. The FCA allows us to visualize how the different classes are distributed in the factorial plane (Figure 4.2). The first two factors summarize 56.75% of the information (factor 1: 30.97% and factor 2: 25.78%). The analysis of the FCA is done by interpreting the axes using opposing data. In our case, the horizontal axis (factor 1) opposes classes 5 and 2, and could correspond to an axis that opposes human biology-related aspects to more evolutionary issues. The vertical axis (factor 2) opposes classes 1 and 2 and may correspond to an opposition between ecological and evolutionary themes. This shows that the theory of evolution belongs to a lexical field isolated from the lexical fields of human biology and, quite surprisingly, from those of ecology and the environment. Classes 3 and 4 are located in the center of the plan and are thus at an intermediate point between the three other classes.

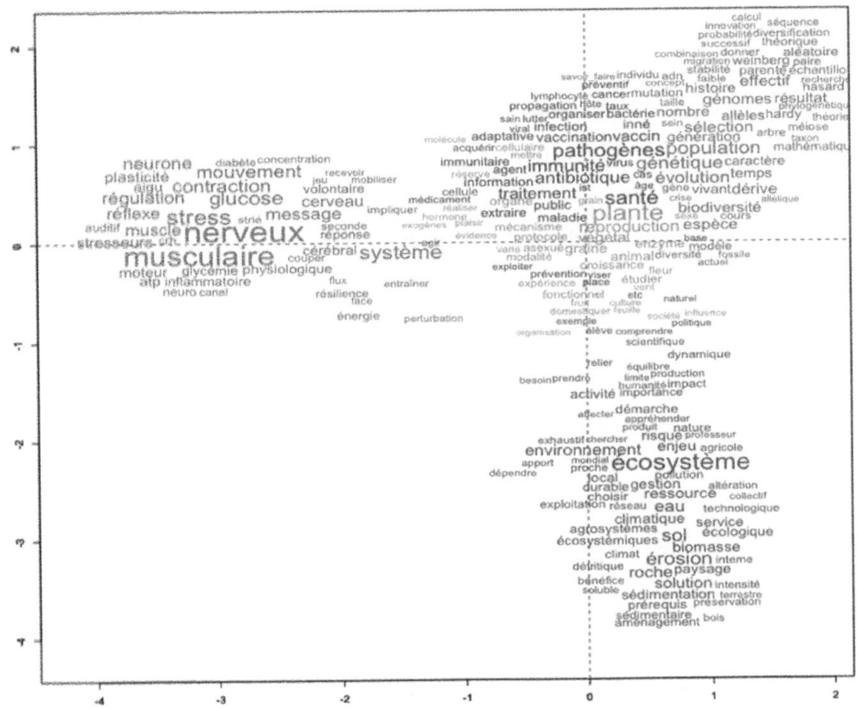

Figure 4.2. *FCA performed on the French biology curriculum set (IRaMuteQ). The text is colored according to Reinert's method classes. For a color version of this figure, see www.iste.co.uk/fortin/evolution.zip*

It appears from the analysis of the official scientific and ELS curriculum that the theory of evolution is well established as it constitutes one of the five classes present in the biology content taught.

In the high school senior class, as part of the common scientific education, a subsection of theme 3 of the school curriculum is entitled "Evolution as a way of reading the world" (*L'évolution comme grille de lecture du monde*).

4.5. Analysis of evolutionary and genetic themes (class 2)

4.5.1. *Overall analysis of the evolutionary and genetic themes*

The analysis carried out allows us to identify which parts of the school curriculum are significantly associated with the evolutionary and genetic themes. This allows us to build a subcorpus containing only the parts of the school curriculum dealing with these themes:

– life and its evolution (7th to 9th grade – ELS);

– the Earth, life and the organization of living things (10th grade – ELS);

– a long history of matter, subsection 1.3 (11th grade – Science);

– the Earth, life and the organization of living things (11th grade – ELS);

– a history of living things, subparts 3.1, 3.2 and 3.3 (high school senior year – science education);

– the Earth, life and the organization of living things (high school senior year – ELS).

The top-down hierarchical classification was carried out on the text segments in a simple way and enabled us to develop five classes (Figure 4.3). The analysis of the five corresponding lexical worlds allows us to propose:

– class A: biodiversity and dynamics of life (dynamic biodiversity, time, etc.);

– class B: transcription and molecular translation (RNA, translation, DNA, transcription, etc.);

– class C: genetics and reproduction (alleles, meiosis, etc.);

– class D: evolutionary classification and human lineage (kinship, homo, classification, etc.);

– class E: population and probabilistic approaches (model, Hardy–Weinberg, estimate, population, etc.).

Classes B and C, which correspond to genetic elements, represent 44.8% of the themes, and evolutionary aspects are present for 55.2%. The first two factors of the FCA summarize 58.74% of the information (factor 1: 33.48% and factor 2: 25.26%). The analysis of the FCA (Figure 4.4) shows that the horizontal axis places class B against classes A, D and E and could correspond to an axis that opposes molecular genetic aspects in favor of more evolutionary questions. The vertical axis puts class

E in opposition to classes A and D, and may correspond to an opposition between the evolutionary themes of classification and biodiversity, and between probabilistic and population approaches. At the factorial level, classes A, C and D are largely superimposed and are distinguished from classes B and E, which belong to other lexical worlds.

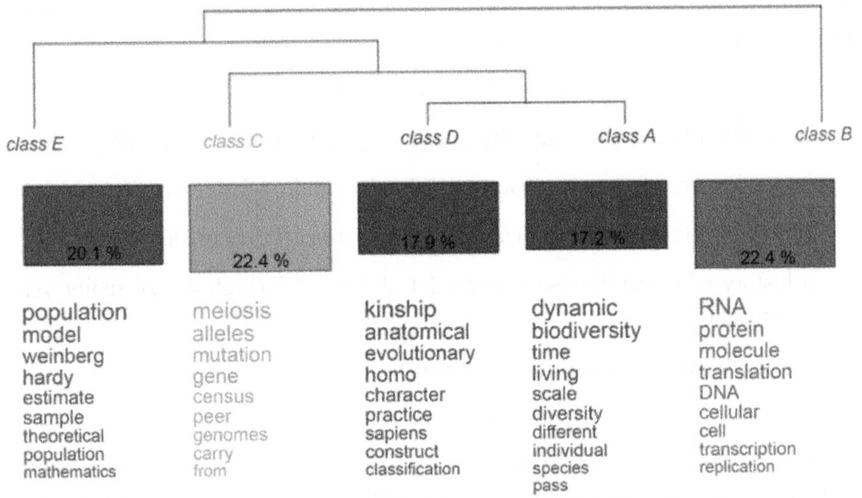

Figure 4.3. *Hierarchical top-down classification using Reinert's method. Carried out on the parts of the biology school curricula that deal with evolutionary and genetic themes (IRaMuTeQ). For a color version of this figure, see www.iste.co.uk/fortin/evolution.zip*

Another interesting aspect of this analysis is evidence of the existence of a population and probabilistic approach in the common science curriculum of the high school senior year. This can be seen in the presence of the Hardy–Weinberg law and the terms "estimate", "sample" and "number", all of which refer to a statistical idea of populations.

In order to better see which parts of the scientific and ELS curricula are associated with classes A, D and E, we used the chi-squared test for association. Table 4.2 shows that classes D and E are mostly related to the new high school senior year science curriculum. This curriculum therefore appears to have some specificity as regards the teaching of evolution. In addition to the overall analyses of the school curricula, we propose, as a second step, a qualitative analysis of the contents of theme 3 of the high school senior year common science curriculum.

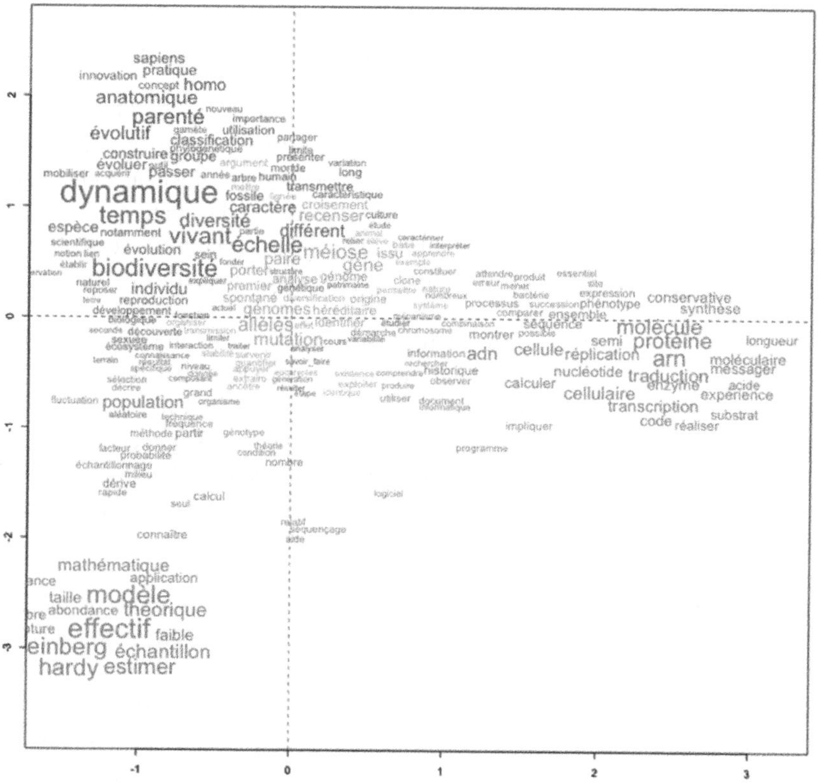

Figure 4.4. *FCA performed on the parts of the French biology school curricula that deal with evolutionary and genetic topics (IRaMuteQ). The text is colored according to the classes of Reinert's method. These words have been collected and displayed in French. For a color version of this figure, see www.iste.co.uk/fortin/evolution.zip*

Levels concerned	Part of the curriculum	Class A	Class D	Class E
7th to 9th grade – ELS	The living world and its evolution	Chi2 = 13.62 p<0.001	Chi2 = 4.14 p = 0.042	NS
10th grade – ELS	The Earth, life and the organization of living things	Chi2 = 24.41 p<0.001	NS	NS
High school senior year – scientific education	Theme 3: a history of life	NS	Chi2 = 16.77 p<0.001	Chi2 = 29.38 p<0.001

Table 4.2. *Chi2 association of the school curriculum with the classes from the lexicometric analysis*

4.5.2. *A look back at the evolutionary themes of the "history of life" theme in high school senior year (BO special edition No. 8 of 2019)*

The section entitled "Biodiversity and its evolution" (*La biodiversité et son évolution*) presents the evolution of populations over time. The presentation of this part specifies that "probabilistic mathematical models and statistical tools make it possible to study the evolutionary mechanisms involved". The content is focused on population dynamics by addressing sampling methods using both the "capture-mark-recapture" method and the Hardy–Weinberg mathematical model[2].

The integration of this mathematical approach is at the service of the construction of a theoretical frequency. Of course, the Hardy–Weinberg law is an idealization, because no real population ever strictly satisfies the conditions that allow its demonstration. In fact, the constancy of theoretical genetic frequencies serves as a reference to understand the deviation from the model and thus show the impact of mutations, migrations, selection and more on the evolution of populations (Gayon and Montévil 2018).

The section "Evolution as a way of reading the world" (*L'évolution comme grille de lecture du monde*) also addresses the explanatory dimension of the theory of evolution: "The concepts of evolutionary biology have a broad explanatory scope, presented here through several examples. They allow us to understand anatomy as the result of a long evolutionary history, made up of adaptations, chance, contingencies and compromises". This section is short (295 words for a science school curriculum that contains 6,650 words, i.e. 4.4%), and therefore contributes only in small part to the lexicometric analysis of the school curriculum. However, its content is interesting, as it clearly identifies the place of biodiversity as a paradigm of biology and proposes the arts school curricula of evolutionary reasoning of anatomical characters. The objective is to show that "Anatomical characters may be the result of natural selection, but some are better explained by the legacy of evolutionary history than by their function". Numerous examples are used to tackle this part, such as the eye, the path of the aortic arch, the male nipple, obstetrical difficulties and wisdom teeth. Evolution is also used as an explanatory model to understand phenomena of medical or agricultural interest. The aim here is to "mobilize evolutionary concepts to explain how microbial populations may, in the long term, no longer be sensitive to a vaccine (or antibiotic) or how the use of phytosanitary products encourages the development of crop pests that are resistant to them".

---

2. This is, in fact, a theorem, which can be proved according to the Mendelian law of segregation under certain conditions. For more details, see Chapter 12.

Finally, the "Human evolution" section (*L'évolution humaine*) deals in a more traditional way with aspects related to human lineage (Quessada and Clément 2013). This part of the school curriculum is more in line with what has already been learned in the previous classes and makes it possible "[...] to present the scientific approach allowing the construction of a reasoned history of human evolution". The work will be based on the construction of a phylogenetic tree based on genetic resemblance and fossil characters to discuss the degree of relationship between *Homo sapiens* and different ape species and the fossils of other *Homo*, Neanderthals and Denisovans. The school curriculum states that "the concepts of relatedness, studied in middle school, are mobilized; particular emphasis is placed on the importance of identifying common evolutionary innovations".

4.6. Discussion and conclusion

At the end of this analysis of the scientific and ELS curricula, we have shown that, on the one hand, the French school curriculum leaves an important place to the theory of evolution from the beginning of secondary school. The analysis carried out on the whole school curriculum shows that the theory of evolution is an important element of the school curriculum in France as it constitutes one of the five main themes of the official scientific and ELS curricula. This confirms the work done on this subject with the old school curricula (Quessada and Clément 2018). As in New Zealand or Greece, French education can be considered to offer a consistent place to evolutionary theory (Campbell and Otrel-Cass 2011; Deniz and Borgerding 2018). However, much of the content in biology – such as aspects related to the human body, physiology and environmental issues – is presented outside of any evolutionary framework. This is also in line with analyses done for other official school curricula, such as in Belgium (Colsoul et al. 2010). This separation of school curricula means that some fields of biology can be taught with a very tenuous link to evolutionary dimensions.

Beyond this incomplete integration, the analysis of the contents concerning the theory of evolution shows the existing links between evolutionary, genetic and dynamic approaches to life. This diversity constitutes an interesting element from a didactic point of view. Indeed, the theory of evolution is often presented as a model or paradigm of historical biology (Orange 2008). This approach, although legitimate, is not necessarily sufficient to ensure a good understanding among the students, and some authors have mentioned the importance of integrating genetics into the teaching of evolution:

> Evolution is a difficult theory for students to understand. Part of the reason for this may be the tendency of instructors to teach evolution in

> the context of ecological systems, isolated from genetic and cellular mechanisms (White et al. 2013, p. 586).

This integration of genetics must not, however, lead to a focus on a small number of explanatory mechanisms (Augusto and Araújo 2020). The French scientific and ELS curricula that address both the genetic bases and those of evolutionary history, as well as approaches to evolutionary dynamics, seem to avoid this pitfall. In this sense, the new science school curriculum for the final year of high school proposes, for aspects of biology, an integration of the theory of evolution as a true explanatory framework that emphasizes population-related and probabilistic approaches.

Our work has led us to a nuanced answer to the question posed by this chapter's title. Although the theory of evolution is largely present in the French school curricula, it would be advisable to develop this integration further so that evolution becomes, at all levels of teaching, the biological "reading grid of the world". If we take Dobzhansky's (1973) proposal seriously, the French school curricula do not yet allow for the full integration of the explanatory power of evolutionary theory. Evolution, although widely present, is not yet the explanatory paradigm for all biology content. In future school curricular developments, the further integration of evolution would give more meaning to learning and perhaps increase appreciation of biology:

> Just as true appreciation for music requires an understanding of how chords, harmony, rhythm, and melody interact, a full understanding and appreciation of the symphony of life demands the theoretical underpinning that only evolution can provide (McComas 2016, p. 91).

4.7. References

Aroua, S., Coquidé, M., Abbès, S. (2012). Controverses dans l'enseignement de l'évolution. Questions de recherches sur les stratégies d'intervention en classe et dans la formation. *Recherches en didactique des sciences et des technologies*, 5, 47–76.

Augusto, L. and Araújo, L. (2020). The centrality of evolution in biology teaching: Towards a pluralistic perspective. *Journal of Biological Education*. doi:10.1080/00219266.2020.1757486.

Barroca-Paccard, M., Orange Ravachol, D., Gouyon, P.H. (2018). Quelle prise en compte du concept de biodiversité dans les programmes français de sciences et technologie et de sciences de la vie et de la Terre ? *Recherches en didactique des sciences et des technologies*, 18, 59–78.

Blancke, S. and Kjærgaard, P.C. (2016). Creationism invades Europe. *Scientific American*, 4(315) [Online]. Available at: https://www.scientificamerican.com/article/eurocreationism/#.

Bybee, R.W. (2002). We should teach about biological evolution. *BioScience*, 52(7), 616–618.

Campbell, A. and Otrel-Cass, K. (2011). Teaching evolution in New Zealand's schools – Reviewing changes in the New Zealand science curriculum. *Research in Science Education*, 41, 441–451.

Colsoul, A., Carette, V., de Biseau, J.-C., Poncelet, J.F., Wolfs, J.L. (2010). L'enseignement de la notion d'évolution biologique en communauté française de Belgique : analyse des savoirs prescrits et des savoirs enseignés. *Actes du congrès de l'Actualité de la recherche en éducation et en formation (AREF)*, Université de Genève, Geneva.

David, P. and Samadi, S. (2011). *La théorie de l'évolution : une logique pour la biologie.* Flammarion, Paris.

Deniz, H. and Borgerding, L.A. (eds) (2018). *Evolution Education Around the Globe.* Springer International Publishing, Cham.

Dobzhansky, T. (1973). Nothing in biology makes sense except in the light of evolution. *The American Biology Teacher*, 35(3), 125–129.

Edwards, J.S. (2002). Dobzhansky's dangerous epigram. *BioScience*, 52(7), 614–615.

Fortin, C. (2009). La théorie de l'évolution : réception et enjeux d'éducation. In *Guide critique de l'évolution*, Lecointre, G. (ed.). Belin, Paris.

Fortin, C. (2014). L'enseignement de l'évolution face aux croyances religieuses. Quelles perspectives curriculaires possibles ? *Histoire, monde et cultures religieuses*, 32, 67–78.

Fortin, C. (2015). Evolutionary theory in secondary schools: Some teaching issues. In *Handbook of Evolutionary Thinking in the Sciences*, Heams Huneman, L. (ed.). Springer, Dordrecht.

Fortin, C. (2018). Le statut épistémique du vivant dans les nouveaux programmes français de la scolarité obligatoire. *Recherches en didactique des sciences et des technologies*, 18, 35–56.

Gayon, J. and Montévil. M. (2018). Répétition et réversibilité dans l'évolution : la génétique des populations théorique. In *Temps de la nature, nature du temps*, Bouton, C. and Huneman, P. (eds). CNRS Editions, Paris.

Griffiths, P.E. (2009). In what sense does "nothing make sense except in the light of evolution?" *Acta Biotheor*, 57, 11–32.

Hermann, R.S. (2008). Evolution as a controversial issue: A review of instructional approaches. *Science et Education*, 17(8–9), 1011–1032.

Hildebrand, D., Bilica, K., Capps, J. (2008). Addressing controversies in science education: A pragmatic approach to evolution education. *Science & Education*, 17(8–9), 1033–1052.

I.A.P, (2006). IAP statement on the teaching of evolution. Resonance. *Journal of Science Education*, 11(11), 93–95.

McComas, W.F. (2016). Biology education is evolution education. *The American Biology Teacher*, 78(2), 91.

Orange, C. (2008). Les fonctions épistémologiques de l'évolution dans les programmes français de sciences de la vie et de la Terre. In *L'évolution du vivant. Un enseignement à risque*, Coquidé, M. and Tirad, S. (eds). Vuibert and SNES-ADAPT, Paris.

Quessada, M.P. and Clément, P. (2007). An epistemological approach to French syllabi on human origins during the 19th and 20th centuries. *Science et Education*, 16, 991–1006.

Quessada, M.P. and Clément, P. (2013). L'évolution humaine dans les programmes et les manuels scolaires de science français de 1994 à aujourd'hui : interactions entre connaissances, valeurs et contexte socioculturel. Report, *Actes du Congrès international AREF*, Montpellier.

Quessada, M.P. and Clément, P. (2018). Evolution education in France: Evolution is widely taught and accepted. In *Evolution Education Around the Globe*, Deniz, H. and Borgerding, L.A. (eds). Springer International, Cham.

Ratinaud, P. (2009). IRaMuTeQ : interface de R pour les analyses multidimensionnelles de textes et de questionnaires [Online]. Available at: http://www.iramuteq.org.

Reinert, A. (1983). Une méthode de classification descendante hiérarchique : application à l'analyse lexicale par contexte. *Cahiers de l'analyse des données*, 8(2), 187–198.

Reinert, A. (1990). Alceste une méthodologie d'analyse des données textuelles et une application : Aurelia De Gerard De Nerval. *Bulletin of Sociological Methodology/Bulletin de méthodologie sociologique*, 26(1), 24–54.

Urgelli, B., Guelladress, K., Quentin, A. (2018). Enseigner l'évolution et la nature des sciences face aux contestations d'élèves : essai de modélisation des postures enseignantes. *Recherches en didactique des sciences et des technologies*, 32, 103–116.

White, P.J.T., Heidemann, M.K., Smith, J.J. (2013). New integrative approach to evolution education. *BioScience*, 63(7), 586–594.

Wouters, A. (2007). Design explanation: Determining the constraints on what can be alive. *Erkenntnis*, 67(1), 65–80.

Academic programs

BO special edition No. 8 of 25 July 2019.

5

What if the Chimpanzee Belonged to the Genus *Homo*? Circulation of Knowledge and Curricular Re-Problematization

Corinne FORTIN

LDAR Université Paris-Cité, Université de Paris-Est-Créteil, France

5.1. Introduction

In France, since the 19th century, the teaching of human evolution, while trying to be as close as possible to current scientific developments, has been caught between science and beliefs (Quessada and Clément 2007; Quessada 2008). As in other countries, we observe that religious beliefs crystallize either opposition or support for the theory of evolution (Reiss 2008, 2011; Long 2012). However, there are also epistemic and cultural components at play (Skoog and Bilica 2002; Alles and Stevenson 2003; Fortin 2009; Sanders and Makotsa 2016). Human evolution thus remains a highly sensitive matter, and its acceptance is not always self-evident due to certain religious or cultural beliefs and cognitive difficulties (Mckeachie et al. 2002; Ingram and Nelson 2006; Hokayem and Boujaoude 2008; Aroua et al. 2012; Pobiner 2016; Kuschmierz et al. 2021).

In the framework of human evolution, the genus *Homo* plays a crucial role; however, in recent years, its definition has been revisited in the light of work in paleoanthropology, molecular biology, genetics and primatology, and new

The Evolution of Life: Teaching, Learning and Training,
coordinated by Corinne FORTIN and Julie GOBERT.

taxonomic proposals have emerged, such as the inclusion of Chimpanzee in the genus *Homo* (Goodman et al. 1998; Page and Goodman 2001; Wildman et al. 2003). In this chapter, we are interested in the circulation of scientific knowledge about the genus *Homo* between high school curricula and university training curricula for teachers of Earth and Life Sciences (ELS).

Insofar as curricula constitute interfaces between the fields of academic research, university training and school teaching, we explore the curricular re-problematization of the genus *Homo* with respect to its scientific problematization. To analyze the training and teaching issues, we draw on the conceptual framework of the curricular didactic approach proposed by Martinand (2016) in connection with curriculum theory (Young 2014; Friesen 2018). Based on the analysis of the curricula of the ELS teacher recruitment competitions and those of secondary education, we question the choices made in the curricular re-problematization of the genus *Homo* through the selection of scientific knowledge and epistemological positions. Then we initiate a reflection on a new curricular approach, more pluralistic, to conceive the genus *Homo* in order to respond to the didactic stakes of its teaching.

Before analyzing the didactic contributions of a pluralist approach to the teaching of the genus *Homo*, it is appropriate to present the conceptual framework of curricular re-problematization on which we rely.

5.2. Conceptual framework of curricular re-problematization in the didactic approach

Depending on the research, curricula instructions are understood as a way of teaching (Jonnaert et al. 2009), an institutional project for the orientation of educational policies, a process for implementing knowledge in the classroom and its evaluation (Audigier et al. 2006), or as "what is supposed to be taught and learned according to a determined order of programming and progression within the framework of a given study cycle" (Forquin 2008, p. 88). The curriculum is therefore not limited to a syllabus or study plan. It is also the manifestation of an educational policy anchored in a sociocultural project (Young 2014). As regards the teaching of evolution in France, educational policies are based on a separation between the religious and scientific registers close to Gould's article, "Non-overlapping magisterial" (2014).

We agree with Chevallard (1997, p. 3) when he claims that there is a national instructional pact built not on an "a priori list of knowledge (…) but on a list of questions on which those who make the pact agree to be educated". The curriculum is thus a socially constructed object, reflecting choices in terms of values and knowledge to be transmitted. Its organization is the result of a selection in response to sociocultural, economic or political aims (Ross 2000) with its implicit, or hidden, component (Apple 1971; Perrenoud 1993). Far from being a homothetic reduction of the scientific *knowledge* into scientific *knowledge* to be-taught or a vertical process of diffusion, of the "top-down" transfer from scientific knowledge towards a school audience, it is the object of a process of translating the "message" when passing from one sphere to another (Callon and Latour 1991). The circulation of scientific knowledge is thus constrained by teaching or training projects (Derouet 2002; Dugal and Leziort 2004), leading to a reconfiguration of scientific knowledge towards a university or school audience that is accompanied by:

– a *didactic transposition* process (Chevallard 1991) resulting from a process of decontextualizing and recontextualizing scientific knowledge into knowledge to be taught, and then into taught knowledge;

– a *curricular transposition* process (Jonnaert 2011) because curricular instructions act as operators of the selection of scientific knowledge to be taught, according to the choices of educational policies;

– a *curricular re-problematization* process (Martinand 2000) because the re-problematization of scientific knowledge to be taught is not of the same nature as scientific problematization. Indeed, re-problematization is part of a process of deconstructing and reconstructing scientific knowledge according to teaching or training objectives in order to make sense of the knowledge "to be taught" and "to be learned".

At the heart of the circulation of knowledge between the scientific sphere, the school teaching sphere and the university teacher training sphere, the curriculum reflects the process of the re-problematization of the genus *Homo* in terms of the aims, selection and elaboration of the content to be taught (Martinand 2005):

– The *aims* are part of the educational policies associated with a societal project. The curriculum is therefore the expression of the choices and decisions made by these policies. It is then a question of determining what place should be given to human evolution and to the genus *Homo* in the recruitment competitions for ELS teachers, which ensure the level of scientific knowledge of the teachers recruited and also include educational and training objectives.

– The *selection* of knowledge to be taught (scientific concepts, students' activities, etc.) is based, on the one hand, on the training objectives assigned to the curricula of the teacher recruitment competitions and, on the other hand, on the educational objectives assigned to school curricula. This is the reason for the interest in exploring the curricular contents of university teacher training in order to understand the dual academic and school construction of knowledge to be taught. This selected knowledge thereby contributes to clarifying the boundaries of the teaching of human evolution and, in particular, those of the genus *Homo* which are given to the pupils, the students and also to the teachers.

– Then, the *elaboration* which corresponds to the insertion of the selected knowledge during the school or university curriculum.

Proposed by Lebeaume (1999), the curricular matrix model also contributes to problematizing the coherence of the knowledge to be taught in the three interrelated poles: the tasks prescribed for the learning activities; the social reference practices used that are associated with the prescribed tasks; and the educational aims that govern the tasks and social reference practices:

– The *educational aims* correspond to the purpose assigned to the acquisition of the knowledge in human evolution in relation to the tasks and their social reference practices.

– *Tasks* are what the student is expected to do. They result in an expected activity to implement acquired knowledge or develop new ones. They thus bring cognitive processes into play in relation to the aims and social reference practices. The task is constrained by the epistemological dimension of the knowledge studied, which determines its purpose as much as the conditions for its realization (Develay 1992).

– *Social reference practices* can be defined as any social activity taken as a reference for school content. As defined by Martinand (2001, p. 19), they correspond to "objective activities of transformation of a natural or human given ('practice'), concerning the whole of a social sector (and not individuals) where the relationship with the proposed activities is not of identity, but in terms of comparison ('reference')". Mobilized within the framework of the tasks and aims, they summon different social spaces relating to the construction of knowledge in human evolution and inscribe the contents taught as socially situated productions.

5.3. The circulation of knowledge relating to the genus *Homo*

In a context of renewal of knowledge on the limits of the genus *Homo*, what are the modalities of knowledge circulation between the fields of scientific research, teacher training and school teaching? Beyond the diffusion or the transmission, what transposition of scientific knowledge is implemented? What educational or training project does the curricula promote in relation to knowledge concerning the genus *Homo*? Is it built from "a list of knowledge *a prior*" or from "a list of questions" to find "an instructional pact" on why and how to teach human evolution?

So many questions that can be answered by analyzing the gaps between scientific problematization and curricular re-problematization. By questioning these gaps, we hypothesize that they are significant of a curricular reconfiguration of the genus *Homo,* mobilizing or excluding certain social reference practices according to the educational or training project. We also hypothesize that the mobilization of certain social reference practices leads the aims of training and education, and consequently drives the circulation of knowledge concerning the genus *Homo*.

The characterization of the genus *Homo* is explored here through a comparative analysis of scientific problematization and curricular re-problematization through the university curricular instructions of the ELS teacher recruitment competitions, and the curricular instructions of ELS secondary education, in order to better understand the selection of the knowledge to be taught. Curricular re-problematization redesigns this scientific knowledge according to the teaching or training project in question, becoming a catalyst for the circulation of knowledge between the scientific sphere, the university sphere and the school sphere (Figure 5.1).

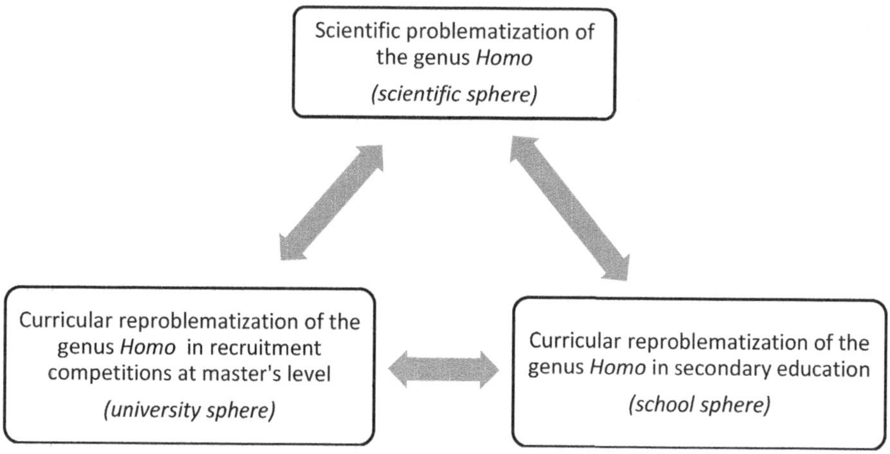

Figure 5.1. *Circulation of knowledge about the genus* Homo *between the scientific, university and school spheres*

5.4. Indicators of the scientific problematization of the genus *Homo*

Even before characterizing the genus *Homo*, the question of characterizing the genus as a supra-specific taxonomic unit arises. Since Linnaeus' *Systema Naturae* (2015), originally published in 1735, binomial nomenclature taxonomy associates, for each organism, a *Genus* name followed by a *Species* name. This is the case for the human species, with its double membership of both the genus *Homo* and the species *sapiens*. Other fossil species share common anatomical attributes with our species and are also included in the genus *Homo*. As a result, in 1864, the first Neanderthal fossil was assigned by King to the genus *Homo* because of anatomical characters close to *Homo sapiens*. Since then, many specimens have been a member of the genus *Homo*, followed by numerous revisions in recent decades.

For example, Simpson (1963) argued, even if the human species and the great apes share a common origin, taxonomy must reflect their divergence with, on the one hand, the genus *Homo* within the Hominid adaptive zone and, on the other hand, the great apes of the present day, which Simpson attached to the Pongid adaptive zone. However, in the 1960s, the discovery of fossils in the Olduvai Gorge contributed to the introduction of new species, such as *Homo habilis*, despite its small cranial capacity of 600 cm^3. According to Leakey et al. (1964), the fossil elements attested to an erect posture, bipedalism and modern dexterity; it was therefore legitimate to include them to the genus *Homo*. However, for some paleoanthropologists (Collard and Wood 2007), because of the numerous anatomical similarities with Australopithecus fossils, the anatomical and adaptive data would rather lead it to be included in the genus *Australopithecus*, and consequently to exclude it from the genus *Homo*.

Cladistics (Hennig 1966) has also contributed profoundly to revisiting systematics and consequently the characterization of the genus *Homo*. For example, the species included in the genus *Homo* should form a monophyletic group and thus form a clade. However, the fossil record is not always sufficient to determine whether these species form a clade or, more broadly, a grade, i.e. common adaptive characteristics; in consequence, they may be poly- or paraphyletic. In this case, how can we classify a species that belongs to a clade while sharing strategies or a common adaptive zone (grade) with other species, without the latter belonging to the same clade? Cladistics then promotes strict monophyly.

Watson et al. (2001) introduced the concept of genetic distance and showed that the distance between chimpanzees and humans is equivalent to the distances between other mammalian species of the same genus. They found that the genetic distances between Gorilla and Chimpanzee, and between Gorilla and Human, are similar to the distances between other mammalian species. For these researchers, the genus *Homo* should also include Chimpanzee, Gorilla and Human. Curnoe and Thorne (2003) confirmed these studies and also concluded that Human and Chimpanzee genomes are sufficiently similar for the two species to be considered congeners. This justifies, according to them, the transfer of the Chimpanzee to the genus *Homo*. In this case, adaptive paleoanthropological criteria, such as bipedalism, the masticatory system or brain volume, are no longer considered a discriminatory factor in belonging to the genus *Homo*. For some biologists, such as Goodman et al. (2001), because of the genetic proximity between the Chimpanzee and our species, the dichotomy between the genus *Pan* and *Homo* is artificial. They propose including the Chimpanzee in the genus *Homo* and to name it the *Homo pan troglodytes* for the common Chimpanzee and *Homo pan paniscus* for the Bonobo. This would induce, de facto, a radical change of the current classification, with the disappearance of fossil genera such as *Australopithecus*, *Paranthropus*, etc., in favor of *H. homo afarenris* instead of *A. afarensis* or *H. homo robustus* instead of *P. robustus*, etc. We would then see a total reconfiguration of the classification of Hominidae, not to say a taxonomic revolution, which some researchers are calling for while others are opposed to it.

But the *genus*, as a supra-specific taxonomic unit, is first and foremost a historical construction that is the result of epistemological orientations, theoretical foundations and different explanatory models that testify to an epistemological pluralism within the scientific community. The contradictory debate animates this pluralism with arguments in favor of or in opposition to the maintenance of certain fossils in the genus *Homo* or the inclusion of the Chimpanzee in the genus *Homo*. Today, four major types of classification are emerging, depending on the characterization of the genus as a taxonomic unit:

– According to Mayr, the *genus* in the evolutionary classification (also called biological) is characterized as that the "genus consists of one species, or a group of species of common ancestry, which differ in a pronounced manner from other groups of species and are separated from them by a decided morphological gap" (Mayr 1950, p. 110) or "a genus (generic taxon) is a monophyletic[1] group composed of one or more species that are separated from other generic taxa by a decided gap" (Mayr and Ashlock 1991, p. 135).

In this case, the *genus* is a group of species with a common origin occupying different adaptive or ecological zones that can be either monophyletic, i.e. including the exclusive common ancestor and all its descendants, or paraphyletic. This classification therefore produces trees that are phylograms.

– Cladistic classification is based on the sharing of synapomorphies and implies a grouping of species that share an exclusive common ancestor. There can therefore be no paraphyly. The genus, then, is a clade constituted by a monophyletic group

1. It is necessary to understand monophyletic in the original sense of the term, i.e. of a common ancestor, and not in the cladistic sense of the exclusive common ancestor and all its descendants.

based on the study of the transformation of plesiomorphous states into apomorphous states of characters. This classification produces cladograms where a *genus* is defined as "a mandatory category to which every species must belong and which contains one species or a monophyletic group of species" (Wiley and Lieberman 2011, p. 318).

– The mixed classification (Collard and Wood 2007) defines a *genus* as "a species, or monophylum, whose members occupy a single adaptive zone". In other words, it would include species that share adaptive strategies with other taxa. According to Wood, for *H. rudolfensis, H. habilis, H. ergaster, H. erectus, H. Heiddelbergensis and H. neanderthalensis* to belong to the genus *Homo*, these species must be closer to *H. sapiens* than to *Autralopithecus*, but phylogenetic analyses show that *H. rulforfensis* and *H. habilis* are detached from the genus *Homo* and are likely to form clades with *Australopithecus*, *Parenthropus*, as well as with *Homo*. In concrete terms, this would lead either to the formation of a new *genus* for these two species, or to their renaming as *A. habilis* and *A. rudolfensis*.

– Phenetic classification emphasizes the link between the degree of similarity and the proximity relationship. However, similarity can be the product of homoplasy (convergence or reversion) and leads either to a paraphyletic group with shared plesiomorphic characters (symplesiomorphies), or to monophyletic groups with shared apomorphic characters (synapomorphies).

The indicators for characterizing the *scientific problematization* of the genus *Homo* are thus initially based on the characterization of the *genus* according to the different scientific classifications, in order to circumscribe it either to the human species and certain close fossil species, or to broaden it to include the great apes (Table 5.1).

Let us recall that the assignment of a fossil or present organism to a genus is the result of a historical scientific construction based on nomenclatural choices according to epistemological choices, such as, for example, taking into account the clade and/or the grade, strict monophyly or paraphyly, etc. The lack of consensus in the scientific community as to how to characterize the genus *Homo* is therefore not only due to a difference in the object of study (fossils vs. molecules) or in the classification choices (phenetic, cladistic, biological), but also to what is inherent to any classification, namely that the taxonomic ranks order, family and genus are not natural entities, but the product of taxonomic rules and of different social reference practices within the scientific community.

Concepts	Reference practice	Type	Genus *Homo*
Homology Ecological niche/adaptive zone Paraphyly Monophyly	Evolutionary/biological classification	Grouping of taxa with different ecological niches, but sharing a common ancestry	Species with a cranial capacity superior to that of australopithecines and having developed a cultural activity
	Phylogram		
Apomorphy Synapomorphy Monophyly	Phylogenetic classification/cladistics	Grouping of taxa belonging exclusively to monophyletic groups (clade and genus overlap)	Extend to Chimpanzee and all fossil descendants belonging to the Chimpanzee–Human clade
	Cladogram		
Apomorphy Synapomorphy Monophyly Ecological niche/adaptive zone	"Mixed" phylogeny classification	Grouping of taxa belonging to monophyletic groups (clade) sharing the same ecological niche (grade)	Exclusion from the genus *Homo of H. habilis* and *H. rudolfensis* moved into the genus *Australopithecus* (*A. habilis, A. rudolfensis*
	Phylo-cladogram		
Overall similarity Distance	Phenetic classification	Grouping of taxa based on the distance or similarity	Expand to Chimpanzee and Gorilla (congeneric species of Humans)
	Phenogram		

Table 5.1. *Elements of characterization of the genus* Homo *according to scientific classifications*

5.5. Indicators of curricular re-problematizations of the genus *Homo*

Curricular re-problematization therefore leads to a rethinking of the teaching of the genus *Homo* according to a teaching or training project based on social reference practices. These can refer to scientific practices or other professional practices or even to those of everyday life. The question of social reference practices also refers to the coherence of the prescribed aims and tasks (Lebeaume 1999; Martinand 2000) in relation to characterizing the genus *Homo* in the curricula.

We are interested, on the one hand, in the curricula of the recruitment competitions for ELS teachers, and, on the other hand, in the curricula of the ELS of secondary education.

5.5.1. *Curricula of competitive examinations for the recruitment of teachers*

In France, recruitment competitions play a key role in the university training of teachers at the master's level. Candidates have access to two competitive examinations, the *Certificat d'Aptitude au Professorat de l'Enseignement Secondaire* (CAPES) to teach ELS in middle and high schools, and the *Agrégation de Sciences de la Vie-Sciences de la Terre et de l'Univers* (SV-STU2) to teach in high schools, as well as selective classes to prepare the entrance for prestigious French *Grandes Écoles*. The main objective of the curricular instructions of these two competitive examinations is to define the scientific knowledge that candidates must master (concepts, technical methods, etc.) in order to certify the required level of qualification.

To explore the curriculum competition exams concerning the genus *Homo*, it is necessary to situate them more globally in terms of aims, structure and selection of knowledge:

– The *aims* are notified in the introduction to the curriculum for the *Agrégation* exam[3], but are also valid for the Capes exam. Knowledge is thus not considered "an end in itself", but as a set of "tools" that give candidates the opportunity to "demonstrate their competences as scientists and future teachers" (p. 1). It is thus a double position, of both scientist and teacher, that is expected.

– The *instructions* are not structured in specific curricula, but are based on the teaching already provided in a bachelor's degree. The two competitive exams have both two components: one in Life Sciences (LS) for the contents of biology, physiology, genetics, ecology, etc., the other in Earth and Universe Sciences (EUS) for the contents in geology, paleontology, the place of the Earth in the universe and the solar system, etc.

– As for the *knowledge selected* in relation to the genus *Homo*, it should be noted that for the *Agrégation* competitive exam, only one item is mentioned "Origin and evolution of Hominidae", cited in the subsection "Fossils, witnesses of evolution" of the EUS part. In the Life sciences part of the Capes[4], the item "Place of the human

2. Sciences de la vie can be translated as Life Sciences and sciences de la terre et de l'univers can be translated as Earth & Universe Sciences.

3. See: https://media.devenirenseignant.gouv.fr/file/agregation_externe_21/92/0/p2021_agreg_ext_svtu_1274920.pdf.

4. See: https://disciplines.ac-toulouse.fr/svt/system/files/202103/CAPES%20external%20SVT%20reference%20biology%202021.pdf.

species in the classification", then the sub-part "Classification of living organisms", establishes several objectives:

> Characterize the relationship of the human species within primates and within the living world by justifying these relationships. Identify the specific characteristics of the human lineage. To justify the construction of the phylogenetic tree of the genus *Homo*. Use genetic and paleogenetic data to reconstruct the evolutionary history of the human species (*Homo sapiens*) and discuss the diversity and current evolution of human populations.

In the EUS reference material for the Capes[5], only one item is mentioned in the subsection "Evolution of life: diversifications and extinctions": "Use paleontological and paleogenomic arguments to reconstruct the major features of the evolutionary history of Hominidae".

Three objectives are thus targeted in the curricula of the two competitions:

– locate the relationship between the human species and other primates;

– present the evolutionary history of the human lineage via the evolution of hominidae;

– build the phylogenetic tree of the genus *Homo*.

By placing the items "Origin and evolution of hominidae" and "Evolutionary history of hominidae" in the field of EUS, the curricula choose to deal only with the paleontological dimension, even though the hominidae family does not have a consensus in the scientific community. It may include Gibbon, Orangutan, Gorilla, Chimpanzee and Human (Page and Goodman 2001; Wildman et al. 2003) or Orangutan, Gorilla, Chimpanzee and Human (Shoshani et al. 1996). Let us recall that this taxonomic pluralism is mainly linked to the contributions of molecular biology and genetics in the construction of phylogenies. Here, however, the circulation of knowledge is only circumscribed to the paleoanthropological dimension of the genus *Homo*.

The curricular re-problematization of the competitions is then centered on the knowledge related to the genus *Homo*, mainly in connection with the social reference practices of phylogenies based on paleoanthropological data. The perspective with molecular phylogenies concerns only present-day hominidae, and

5. https://disciplines.ac-toulouse.fr/svt/sites/svt.disciplines.ac-toulouse.fr/files/capes_external_and_cafep/capes_external_svt_-_2020_-_geology_reference.pdf.

not the characterization of the genus *Homo*. We note, however, a tension between the items of the *Agrégation* competition centered on paleoanthropology (EUS part) and the items of the reference material of the Capes competition, with the consideration of the genus *Homo* and its history in both the LS and EUS parts.

Indicators			Aim of training	Tasks
Concepts	Reference practices	Genus *Homo*		
Apomorphies Common ancestor	Phylogeny based on paleoanthropological and paleogenetic data	The current human species and its closest fossil relatives	Certification (phylogeny of hominidae)	Analysis of fossil and paleogenetic data to establish relationships

Table 5.2. *Elements of curricular re-problematization of the genus* Homo *of recruitment competitions*

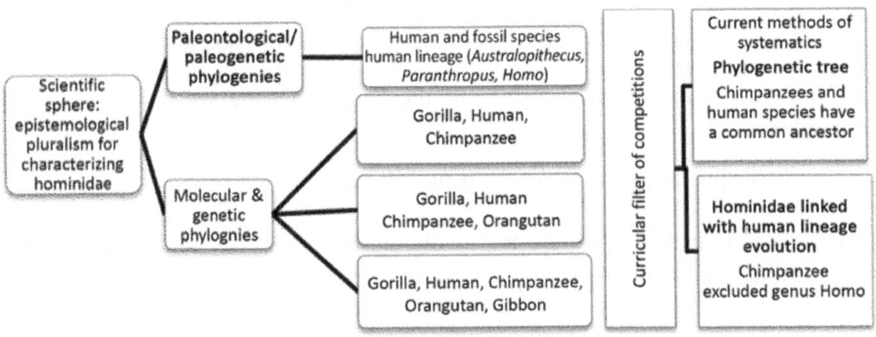

Figure 5.2. *Selection of the paleoanthropological approach to the genus* Homo *(in bold) in the curricula of recruitment competitions*

In summary, the epistemological pluralism as well as the dissensus expressed within the scientific community to characterize the limits of hominidae and in particular those of the genus *Homo* are not mentioned in the competitions curricula. Only possible "technical" difficulties encountered in the construction of phylogenies are mentioned.

In a context where the place of the human species in the evolutionary history is not always culturally accepted, it is surprising that at the master's level, the critical dimension relating to the limits of the genus *Homo* is not a training issue. Pluralist approach would avoid a form of essentialization of the genus *Homo*, or even the risk of reactivating a Simpsonian approach to Hominidae among teachers in training. While it is always possible, during training, to sensitize future teachers to critical and pluralistic approaches to the genus *Homo*, this approach, which is left solely to the initiative and discretion of the trainers, remains marginal in view of the density of curricular instructions. Only the items explicitly mentioned for the competitions will be given priority.

It should also be noted that the curricular re-problematization does not call upon the cognitive difficulties and epistemological or sociocultural obstacles that future teachers are likely to encounter among their students (misconceptions, religious beliefs, etc.). Only a reminder of the history of science is mentioned (creationism, fixism, Lamarckism, Darwinism, etc.) to illustrate the construction of evolutionary thought. Consequently, the curricula of the competitions reflect a training project that does not explicitly question the classification choices concerning the limits of the genus *Homo,* such as the possible grouping of chimpanzees and humans in the same genus, if only to maintain a unity of taxonomic rank.

5.5.2. *School curricula*

In order to situate the curricular stakes of teaching and learning about the genus *Homo* in secondary education in terms of *aims*, *structuring* and *selection* of knowledge, the indicators are to be sought in the *Bulletin Officiel du Ministère de l'Éducation Nationale* (Official Bulletin of the French Ministry of Education), ensuring a common culture shared by all students.

The *aims* specifically focused on in the teaching of evolution are based on two important prerequisites: "evolution is a grid for reading the world", and evolution "constitutes a powerful tool for understanding the living world" (BO no. 8, 2019, 12th grade, *Terminale Scientifique*).

The teaching is structured in two stages: a first one, in middle school (BO no. 30, 2018 and BO 2020) and a second one, in high school (BO no. 8, 2019). In both cases, the genus *Homo* is linked to *H. sapiens* and to fossil species *Homo.* In middle school, in the topic "Life and its evolution", the item "Linking the study of relationships between living beings and evolution" is broken down into two sub-items: "Shared characters and classification" and "The major groups of living beings, including *Homo sapiens*, their relationships and their evolution".

In high school, in the topic "A History of Living Being" and in the subsection "Human Evolution", it is specified that the objective "is not to conduct an exhaustive approach to fossils species and their biological characteristics, but to present the scientific approach to building a "reasoned history" of human evolution "based on fossil records".

Concerning the selection of knowledge, in middle school, the focus is on evolutionary relationships. The Eduscol[6] supplements specify that "all present and fossil species have a degree of relationship that is more or less important according to the evolutionary innovations they share". It is therefore a question of exploring "the degree of evolutionary relationship of present-day and/or fossil organisms on the basis of shared characters" and of taking into account "the evolutionary history of large groups of living beings, including *Homo sapiens*" in order to establish an evolutionary relationship. However, no information on characterizing the genus *Homo* is provided.

The year 2019 was also a pivotal one, between the transitional maintenance of previous curricula and the implementation of new ones in 2020. The corpus relating to the re-problematization of the genus *Homo* is based, here, on the curricular instructions prior to 2019 (BO no. 8, 2011), as well as on those from 2019 onwards (BO no. 1 and no. 8, 2019) to better identify both their continuities and their discontinuities. The instructions in effect until 2019, in the high school senior science class (12th grade), were centered on "The place of Man in the classification: evolutionary relationship of Man within the primates and the human lineage". They insisted on the diagnosis of the genus *Homo* in order to situate present or fossil primate species in a phylogenetic tree. The genetic and chromosomal comparison between the human species and the chimpanzee was also mentioned. The item "The place of Man in the classification" was broken down into three sub-items: "characterize the evolutionary relationship within the primates; characterize the human lineage; argue the of the phylogenetic tree building of the genus *Homo*". On this last point, the instructions invited caution, by discussing, in particular, the different possible evolutionary scenarios. Another section was devoted to the diversification of living beings without genome modifications, including the cultural transmission of behaviors with possible examples in primates, but without any link with a possible phylogeny of behaviors.

The new curricular instructions, in effect from 2020, take up the relationships between present-day species and the fossils of Hominidae through phylogeny, by pointing out the fossil discoveries made in paleontology in the *Terminale* class (12th grade, 18-year-olds). In *Première* class (11th grade, 17-year-olds), the contributions of the study of the human genome are mentioned in order to retrace a part of its history. As for human evolution, paleoanthropology is called upon to develop "a

6. Eduscol is a website run by the French Ministry of Education that provides information and resources for teachers.

scientific narrative of our origins based on the fossil record. Phylogeny allows for the study of evolutionary relationships between present and fossil species of Hominidae" (BO no. 8, 2019). These new instructions also emphasize the development of cranial capacity as an important element in the diagnosis of the genus *Homo*, and the relationship between *H. sapiens*, Neanderthals and Denisovans. The reference to classification is made through the sharing of characters or evolutionary innovations. As for the phylogenetic links between the human species and the other great apes, it is specified that "the present human species (*H. sapiens*) is part of the primate group and is more particularly related to the great apes with whom it shares morpho-anatomical characters and genetic similarities". It is also specified that "it is with the Chimpanzee that it shares the most recent common ancestor". Finally, a recommendation is made to "relate the genetic similarity between primate species and their degree of evolutionary relationship".

In the *Seconde* class (10th grade, 16-year-olds), we find an item that was present in the previous science curriculum, namely, "in some animals, acquired behaviors can be transmitted from one generation to the next and constitute a source of diversity" (BO no. 8, 2019), thus underlining the cultural dimension of some behaviors in primates, with this time the idea of "apprehending the notion of cultural evolution and its links with that of biological evolution" in the final year of secondary school, but without making an explicit phylogenetic link.

The previous and current high school curricula are broadly in continuity (except for the addition of paleogenetics). Both refer to the "human lineage" to emphasize that the fossil species of this lineage are more closely related to the genus *Homo* than to the Chimpanzee. However, the ambiguity of the phrase "human lineage" may suggest a form of phyletic gradualism of which our species would be the achievement. Thus, there is a difference in the detail of the diagnosis of the genus *Homo*. In the previous curriculum, it was indicated that this genus included "present-day humans and a few fossils species that are characterized by a reduced facial prognathism, little sexual dimorphism on the skeleton, a style of bipedalism with an advanced foramen magnum and an aptitude for running, a parabolic mandible, etc. The production of complex tools and a variety of cultural practices are also associated with the genus *Homo*, albeit not exclusively. The accurate building of the phylogenetic tree of the genus *Homo* is controversial in detail" (BO no. 1, 2019). In the current curriculum, the characterization of the genus *Homo* is more succinct. Its diagnosis is essentially summarized by morpho-anatomical criteria: increase in brain volume, prolonged bipedalism and the shape of the mandible.

The notion of an evolutionary relationship, introduced in middle school, is thus remobilized in high school. Emphasis is placed on the importance of identifying "common evolutionary innovations". The paleoanthropological approach associated with paleogenetics (e.g. Neanderthal and Denisovan heritage) inscribes the history of human evolution in connection with these fossil discoveries.

<table>
<tr><th rowspan="2"></th><th colspan="3">Indicators</th><th rowspan="2">Mentions</th><th rowspan="2">Tasks</th></tr>
<tr><th>Concepts</th><th>Practices</th><th>Genus Homo</th></tr>
<tr><td>Middle school</td><td rowspan="3">Sharing of innovative features (apomorphy)
Common ancestor</td><td rowspan="2">Phylogenetic classification</td><td>No diagnosis</td><td rowspan="2">Scientific literacy on the relationship between present and fossil hominidae species</td><td>Compare characters
To argue the evolutionary history of H. sapiens</td></tr>
<tr><td rowspan="2">High school</td><td>The human species and fossil species marked by a development of the cranial capacity</td><td>Compare anatomical characteristics and genetic similarities among primates (degree of evolutionary relationship and building of a phylogenetic tree)</td></tr>
<tr><td>Tracing evolutionary history</td><td></td><td>Construction of a scientific account of the origin of the human species from fossil data</td><td>Position fossil species in a phylogenetic tree, analyze scientific arguments to clarify the relationship of H. sapiens with other Homo (Neanderthals and Denisovans)</td></tr>
</table>

Table 5.3. *Elements of curricular re-problematization of the genus* Homo *in middle and high schools*

The curricular re-problematization of the genus *Homo*, in middle and high schools, follows the curricular orientations of the recruitment competitions for ELS teachers. It uses phylogenetic classification to highlight the links of the relationship between the present and fossil species on the basis of genetic, morpho-anatomical, paleontological and paleogenetic data. This re-problematization is mainly centered on human lineage and not on the characterization of the genus *Homo*, and it is based on social reference practices from paleoanthropology and paleogenetics.

5.6. Circulation of knowledge and re-problematization of the genus *Homo*

The circulation of knowledge about the genus *Homo* between the three spheres (scientific, university training and school education) is part of a process of institutionalization of knowledge. But what is really circulating? Mainly, a *paleoanthropological approach* to the genus *Homo* with its objects of investigation: the fossils; its practices of morpho-anatomical and paleogenetic analysis; and its phylogenetic productions. The other classification approaches are not used because the curricular re-problematizations promote the operativity of the genus *Homo* in the human lineage.

The gaps observed between the three spheres thus result from a dissociation between the knowledge and the social production practices in which they take on meaning. Consequently, the circulation is accompanied by a re-problematization by introducing tension between the various fields.

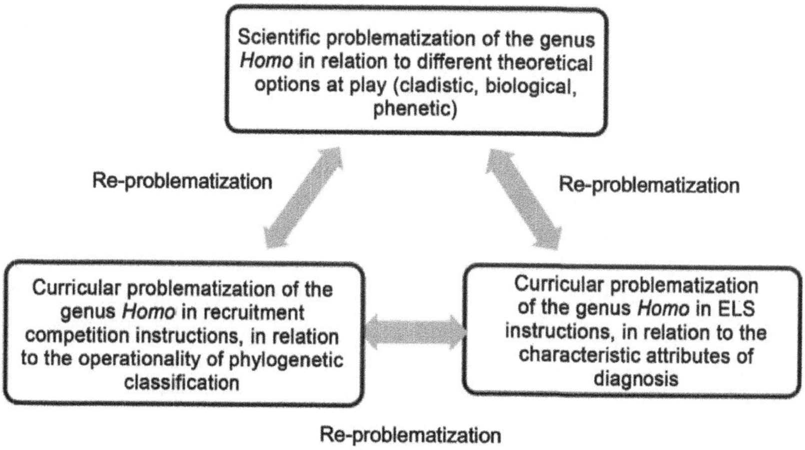

Figure 5.3. *Circulation and gaps between the three problematization fields of the genus* Homo

The gap between the scientific characterization of the genus *Homo* and its characterization in the curricula of competitive examinations results in a tension between a pluralist approach in relation to the different theoretical options of classification and a focus on fossil Hominidae in relation to human evolution. The genus *Homo* then finds its place in the field of paleoanthropology, as an operator, to restore human evolution. However, the limits of the genus *Homo* are not questioned, as, for example, in the case of *Homo habilis*, which could also be classified as an australopithecine.

The gap between the scientific characterization of the genus *Homo* and its characterization in the ELS school curricula results in a tension between a phylogenetic approach of sharing characters and/or evolutionary innovations and a typological approach based on its diagnosis in order to position it in a phylogenetic tree.

Last point, the gap between the characterization of the genus *Homo* in the competition curricula and the school curricula of ELS is not a tension, but rather a transposition of the items present in the competition curricula into the school curricula, oscillating between the place of the genus *Homo* in the phylogeny of primates and that of the human lineage.

The curricular dimension of the genus *Homo* is not so much that of an update of scientific knowledge as the congruence between an educational or training project and the social reference practices mobilized. The preferred angle in the curricula remains the hominization process, even if the term is not used. It is mainly a question of marking the divergence from the other great apes on the basis of a *narrative* of human evolution that is rebuilt as paleontological and paleogenetic discoveries are made. As a consequence, the great apes are still often described in a subtractive way of thinking, without the attributes that make the human being unique. Thus, school curricula focuses on the Chimpanzee as an extra-group to emphasize the relationship between fossil *Homo* and *H. sapiens sapiens*, and not to emphasize the relationship between *Pan* and *Homo*. This results in a curricular re-problematization that can reinforce the idea of an irreducible duality between the two species, because only plesiomorphic morpho-anatomic characters of the Chimpanzee and apomorphic characters of the human species are then compared. This has the effect of provoking a *cognitive dissonance* as to a common origin based on the sharing of an apomorphic character, since, on the one hand, plesiomorphic characters are indicated (e.g. reduced cranial volume, eccentric foramen, large canines, opposable toes, two vertebral curvatures, long and narrow pelvis, etc.) as if the Chimpanzee did not present apomorphic characters, and on the other side plesiomorphic characters (e.g. developed cranial volume, centered foramen, small canines, four vertebral curvatures, short and wide pelvis, etc.), as if the human species did not present plesiomorphic characters.

Emphasis is also placed on the specificity of humans without really discussing the attributes of the human lineage. Nothing is said about what makes the Chimpanzee unique, except the emphasis on the plesiomorphic characters compared to the apomorphic characters of the Human, at the risk of implicitly presenting our species as more evolved than the other great apes, even if this is not the objective of the curricula.

For example, the question of language is not mentioned. Yet, the acquisition of symbolic language is also possible in great apes, even in the absence of articulated language. As shown in the cases by the female chimpanzee Washoe, who learned sign language and passed it on to her descendants[7], and the bonobo Kanzi who, for years, communicated with researchers using lexigrams. Similarly, the production of lithic tools is not a characteristic specific to *Homo*, since we know of tools dated between 2.6 and 2.3 million years ago in East Africa, and that australopithecines probably also used tools. Lastly, as for permanent bipedalism, if it is an effective characteristic of the genus *Homo*, it is only a specialization of a locomotor register among others.

The curricular re-problematization thus circulates a narrative about the genus *Homo* that is documented by fossil remains without ever questioning its limits and the possible inclusion of the Chimpanzee in this genus. This is undoubtedly because the sociocultural behaviors of the great apes are treated solely from an ethological perspective, independently of evolution. However, chimpanzees, like humans, produce and use tools, have social organizations and communicate through a conceptual language. The origin of social and cultural behaviors is therefore not integrated into phylogenetic building, as advocated by some researchers (Duda and Zrzavy 2013). Such an approach assumes that we share sociocultural processes with other great apes which can be as much a lever as a major obstacle to accepting the idea that we share a common ancestor with the Chimpanzee (Fortin 2009, 2015).

In short, the curricular re-problematizations do not or hardly question *the sealing or porosity of the genus Homo* with the genus *Pan*. This choice is, moreover, consistent, insofar as it allows both students and future teachers to access paleontological data, but it can also be a limiting factor in understanding the epistemological building of the genus *Homo*, in particular, for teachers in university training. Indeed, with the circulation of knowledge being limited mainly to fossil records, the question of the possible belonging of the Chimpanzee to the genus *Homo* is still not on the agenda.

7. Chimpanzees are capable, in a human context, of learning a human language.

These re-problematizations have the advantage of "fleshing out" the genus *Homo* while running the risk that the inference of the diagnosis on phylogeny will lead to its naturalization or even its essentialization. This leads to a *circularity* rather than a circulation of knowledge: if all the attributes of the diagnosis define the genus *Homo*, these attributes are necessarily understood as being unique to the human. Through this "nominalistic" approach to the genus *Homo*, is there not a risk of opening the door to a rationalization of prejudices about the human exception by reinforcing an anthropocentric training and education project?

5.7. To renew the circulation of knowledge: new ways of thinking about curricular re-problematization

At present, curricula characterize the genus *Homo* without indicating how it has been epistemologically built and is still being built today. If it is understandable that at the school level, the question of the limits of the genus *Homo* is not dealt with because of its complexity, it is more surprising that the curricula of the recruitment competitions, at the master's level, do not explicitly refer to it. By dismissing from the outset the epistemological pluralism within the scientific community with, for example, the possibility that humans and chimpanzees are in the same taxonomic unit of genus, the curricula of the competitive examinations give future teachers a reified image of the genus *Homo*. It is therefore important to circulate data (paleontological, genetic, behavioral, paleo-climatological, paleo-ecological, etc.), classification methods (cladistic, biological, phenetic) and evolutionary models (origin and expansion *of H. sapiens*) by re-problematizing them in a training perspective based around the need for contradictory debate within the scientific community (e.g. *habilis* as *Homo* or *Australopithecus*, or the inclusion of the chimpanzee in the genus *Homo*, etc.).

The goal is to make trainee teachers realize that the validity of arguments for or against the inclusion of the Chimpanzee in the genus *Homo* or *habilis* in the genus *Australopithecus* depends as much on fossil, genetic and behavioral data as it does on choices of prioritization between phylogeny and adaptive strategies. For example, choices to assign *habilis* to the genus *Homo* or *Australopithecus* and Chimpanzee to the genus *Pan* or *Homo* have an impact on the interpretation of the entire fossil record. In other words, the nature of the data, depending on the social reference practices used, shapes our interpretation of a given genus, which is why it is important for future teachers to be aware of this. However, two pitfalls must be avoided: reductionism and relativism. The first would lead to an intangible definition of the genus *Homo*, while the second would suggest that the characterization of the genus *Homo* is merely arbitrary.

Knowing that the taxonomic status of the genus *Homo* is not fixed, but rather continually discussed and debated within the scientific community, a question arises: should epistemological pluralism be integrated into the curriculum or not, and if so, how should it be integrated into university training and/or school education? Depending on the answer chosen, two curricular orientations emerge:

– the first is that not integrating it is the current curricular approach;

– the second option seems essential to us. It is based on the taxonomic flexibility of the genus *Homo*, which, since Linnaeus, has been constantly re-examining our relationship with the other great apes, or as Leakey said after having taken note of Goodall's observations on the production and use of tools in chimpanzees: "now we must redefine tool, redefine Man, or accept chimpanzee as humans" (Goodall 1998, p. 2084).

Whether for university teacher training or for secondary education, it is desirable to go beyond a naive approach to scientific production according to which the data (paleontological or molecular) would speak for themselves outside of any prior explanatory field (Fortin 2014, 2015). The characterization of the genus *Homo* also provides access to a community of scientific practices with the validation processes that are at work. Understanding, for example, that the "locomotion" of the genus *Homo* from the time of Linnaeus (bipedalism and quadrupedalism) to the present day (only bipdedalism) is not a chaotic path, but rather has a heuristic function that allows us to rethink and update the place of our species among the present and fossil species of great apes.

Consequently, we propose that curricular re-problematization should not be limited to a "monstration" of the genus *Homo* by exhibiting anatomical or molecular data, but should also be organized around epistemological pluralism in a critical approach that takes into account the question of our relationship of otherness to other great apes (Zeitoun 2015; Lequin 2018). This implies articulating research in paleo-anthropology, molecular biology and primatology, in an integrative approach rather than by maintaining the current dichotomy between Earth Sciences, on the one hand, and Life Sciences, on the other hand.

However, recognizing the human species as a great ape united by phylogenetic proximity to other great apes is not necessarily self-evident. Despite the Darwinian revolution, there is still a reluctance, even a resistance, outside the scientific community, to conceive of chimpanzees and humans as current representatives of two evolutionary branches from a common ancestor (Fortin 2009, 2016; Wagner-Egger et al. 2018). Even among teachers, Silva et al. (2018) note the difficulty of accepting the idea that the chimpanzee may eventually belong to the genus *Homo* because this change would make a demarcation between our species and other great apes porous.

Thinking about our relationship with the great apes means understanding that our phylogenetic proximity is not only morpho-anatomical, genetic or molecular. It is also behavioral and, more broadly, cultural. Today, we know that we share, in addition to morpho-anatomical, molecular and genetic attributes, a common psychic activity based on the perception of the consciousness of the self and others (Anderson 2003; Vauclair and Cochet 2013). A new look is thus taken upon these great apes by reconfiguring the limits that were previously assigned to them. From now on, they are no longer only wild beings, but also beings using and modifying their environment to feed themselves, build a habitat or ensure the protection of their young.

Without entering, here, into a necessarily restrictive definition, *culture* can nevertheless be characterized without our species being the measure of it. If we consider that socially transmitted information generates behaviors within a population of a given species of great apes, and that this information is not necessarily the same within different populations of the same species, then behaviors may be different from one population to another, and produce cultural variations qualified as cultural "traditions" (Whiten 2017). This is therefore not an exclusive property of the genus *Homo*, but rather a shared continuum among great apes in connection with our phylogenetic proximity (Laland et al. 2009). The divide between *nature* and *culture* is thus called into question, because the great non-human primates are not exclusively the product of genetically determined behaviors, but also of cultural transmissions. The evolution of great primates is now understood as a co-evolution between gene and culture with the production of abstract cognitive reasoning (Boyd and Richarson 1985; Laland and Janik 2006; Rey et al. 2018). This is not, however, to deny the differences between human and non-human primates, nor to naturalize or biologize culture, or to go beyond an exclusively anthropocentric approach to culture. There is no need to oppose the great non-human primates to humans, nor to hierarchize them, but only to recognize them both in their singularity and in their diversity. For humans, as for chimpanzees, behavior and social life present similarities, as well as differences resulting both from a common evolutionary history and from their specific evolution.

5.8. Conclusion

The circulation of knowledge concerning the genus *Homo* necessarily has a scientific and cultural impact. The choice to privilege one classification over another is not only an internal question for the scientific community, but is also a sociocultural issue. This perspective makes it possible to refocus on the educational purposes of the scholarly knowledge to be taught, as well as on the educational or training experiences to be provided to students and teachers. Based on the analysis of curricula, it seems necessary to open new curricular perspectives to facilitate the circulation of knowledge by engaging in new curricular re-problematizations on the genus *Homo* between the scientific sphere and the fields of university training and secondary education.

The curriculum cannot be limited to a tool for programming teaching and learning content of knowledge; it also aims to keep epistemological pluralism alive as a driving force for the circulation of knowledge. Indeed, the knowledge that circulates is not necessarily stabilized, as shown by the still open questions of the inclusion of the chimpanzee in the genus *Homo* or of *Habilis* in the genus *Australopithecus*. It is not a question of considering these new proposals right or wrong, but of understanding that both respond to different epistemological concerns. Questioning the epistemological pluralism of the limits of the genus *Homo* then becomes an issue of training for teachers and education for students. This leads us to rethink the curricula of training and education with a critical view of the circulation of knowledge because the data (anatomical, molecular, genetic, etc.) do not speak for themselves. Under these conditions, the question of the delimitation of the genus *Homo* becomes an object of training and teaching that must be re-problematized in the light of the epistemic and cultural issues it carries.

5.9. References

Alles, D.L. and Stevenson, J.C. (2003). Teaching human evolution. *The American Biology Teacher*, 65(5), 333–339.

Anderson, J.R. (2003). De l'autre côté du miroir. À la recherche de la reconnaissance de soi chez les singes. In *Aux Origines de l'humanité, Volume 2 : Le propre de l'homme*, Picq, P. and Coppens, Y. (eds). Fayard, Paris.

Apple, M.W. (1971). The hidden curriculum and the nature of conflict. *Interchange*, 2(4), 27–40

Aroua, S., Coquidé, M., Abbes, S. (2012). Controverses dans l'enseignement de l'évolution. Questions de recherches sur les stratégies d'intervention en classe et dans la formation. *RDST – Recherches en didactique des sciences et des technologies*, 5, 47–76.

Audigier, F., Crahay, M., Dolz, J. (2006). *Curriculum, enseignement et pilotage*. De Boeck Supérieur, Brussels.

Boyd, R. and Richerson, P.J. (1985). *Culture and the Evolutionary Process*. University of Chicago Press, IL.

Callon, M. and Latour, B. (1991). *La science telle qu'elle se fait. Anthologie de la sociologie des sciences de langue anglaise.* La Découverte, Paris.

Chevallard, Y. (1991). La transposition didactique : du savoir savant au savoir enseigné. In *La Pensée Sauvage*, 2nd edition, Strauss, C.-L. (ed.). Pocket, Grenoble.

Chevallard, Y. (1997). Questions vives, savoirs moribonds : le problème curriculaire aujourd'hui. *Actes du colloque Défendre et transformer l'école pour tous*, 3–5.

Collard, M. and Wood, B.A. (2007). Defining the genus *Homo*. *Handbook of Paleoanthropology*, 3, 1575–1611.

Curnoe, D. and Thorne, A. (2003). Number of ancestral human species: A molecular perspective. *Homo*, 53(3), 201–224.

Derouet, J.L. (2002). Du transfert à la circulation des savoirs et à la reproblématisation. De la circulation des savoirs à la constitution d'un forum hybride et de pôles de compétences. Un itinéraire de recherche. *Recherche & formation*, 40(1), 13–25.

Develay, M. (1992). *De l'apprentissage à l'enseignement.* ESF, Paris.

Duda, P. and Zrzavý, J. (2013). Evolution of life history and behavior in Hominidae: Towards phylogenetic reconstruction of the chimpanzee–human last common ancestor. *Journal of Human Evolution*, 65(4), 424–446.

Dugal, J.P. and Léziort, Y. (2004). La circulation des savoirs entre recherche et formation : l'exemple des concepts didactiques lors d'une action de formation de conseillers pédagogiques. *Revue française de pédagogie*, 37–47.

Forquin, J.-C. (2008). *Sociologie du curriculum.* PUR, Rennes.

Fortin, C. (2009). *L'évolution à l'école : créationnisme contre darwinisme ?* Armand Colin, Nantes.

Fortin, C. (2014). L'enseignement de l'évolution face aux croyances religieuses. *Histoire, monde et cultures religieuses*, 4, 67–78.

Fortin, C. (2015). Evolutionary theory in secondary schools: Some teaching issues. *Handbook of Evolutionary Thinking in the Sciences*, 897–910.

Fortin, C. and Reier-Røberg, K. (2016). De la nature de la causalité évolutive à la nature de la biologie évolutive : types de causalité mobilisés par les élèves en classe de seconde. *RDST – Recherches en didactique des sciences et des technologies*, 14, 125–152.

Friesen, N. (2018). Continuing the dialogue: Curriculum, Didaktik and theories of knowledge. *Journal of Curriculum Studies*, 50(6), 724–732.

Goodall, J. (1998). Learning from the chimpanzees: A message humans can understand. *Science*, 282, 2184–2185.

Goodman, M., Porter, C.A., Czelusniak, J., Page, S.L., Schneider, H., Shoshani, J., Gunnell, G., Groves, C.P. (1998). Toward a phylogenetic classification of primates based on DNA evidence complemented by fossil evidence. *Molecular Phylogenetics and Evolution*, 9(3), 585–598.

Goodman M., Czelusniak J., Page S.L., Meireles C.M. (2001). Where DNA sequences place *Homo sapiens* in a phylogenetic classification of primates. In *Humanity from African Naissance to Coming Millenia*, Tobias, P.V., Raath, M.A., Moggi-Cecchi, J., Doyle, G.A. (eds). Firenze University Press, Florence.

Gould, S.J. (2014). Non-overlapping magisteria. *Filozoficzne Aspekty Genezy*, 11, 7–21.

Hennig, W. (1966). *Phylogenetic Systematics*. University of Illinois Press, Urbana, IL.

Hokayem, H. and Boujaoude, S. (2008). College students' perceptions of the theory of evolution. *Journal of Research in Science Teaching*, 45(4), 395–419.

Ingram E.-L. and Nelson C.-E. (2006). Relationship between achievement and students' acceptance of evolution and creation in an upper-level evolution course. *Journal of Research in Science Teaching*, 43(1), 7–24.

Jonnaert, P. (2011). Curriculum, entre modèle rationnel et irrationalité des sociétés. *Revue internationale d'éducation de Sèvres*, 56, 135–145.

Jonnaert, P., Ettayebi, M., Defise, R. (2009). *Curriculum et compétences : un cadre opérationnel*. De Boeck Supérieur, Louvain-la-Neuve.

Kuschmierz, P., Beniermann, A., Bergmann, A., Pinxten, R., Aivelo, T., Berniak-Woźny, J., Fortin, C., Graf, D. (2021). European first-year university students accept evolution but lack substantial knowledge about it: A standardized European cross-country assessment. *Evolution: Education and Outreach*, 14(1), 1–22.

Laland, K.N. and Janik, V.M. (2006). The animal cultures debate. *Trends in Ecology and Evolution*, 21(10), 542–547.

Laland, K.N., Kendal, J.R., Kendal, R.L. (2009). Animal culture: Problems and solutions. *The Question of Animal Culture*, 174–197.

Leakey, L.S.B., Tobias, P.V., Napier, J.R. (1964). A new species of the genus *Homo* from Olduvai Gorge. *Nature*, 202, 7–9.

Lebeaume, J. (1999). Perspectives curriculaires en éducation technologique. Mémoire d'habilitation à diriger des recherches, Université Paris Sud.

Lequin, M. (2018). La révolution paléobiologique at-elle eu lieu en paléoanthropologie ? *Lato Sensu, revue de la Société de philosophie des sciences*, 5(1), 26–36.

Linneaus, C. (2015). *Systema Naturae*. Andesite Press.

Long, D.E. (2012). The politics of teaching evolution, science education standards, and being a creationist. *Journal of Research in Science Teaching*, 49(1), 122–139.

Martinand, J.-L. (2000). Production, circulation et reproblématisation des savoirs. Communication au Colloque international de sciences de l'éducation tes pratiques dans l'Enseignement Supérieur. Université de Toulouse le Mirail.

Martinand, J.-L. (2001). Pratiques de référence et problématique de la référence curriculaire. In *Didactique des disciplines, les références au savoir*, Terrisse, A. (ed.). De Boeck, Brussels.

Martinand, J.-L. (2005). L'éducation technologique à l'école moyenne en France : problèmes de didactique curriculaire. *La revue canadienne de l'enseignement des sciences mathématiques et des technologies*, 3–1, 49–63.

Martinand, J.-L. (2016). La question de la référence en didactique du curriculum. *Investigações em Ensino de Ciências*, 8(2), 125–130.

Mayr, E. (1950). Taxonomic categories in fossil hominids. In *Cold Spring Harbor Symposia on Quantitative Biology*. Cold Spring Harbor Laboratory Press, New York.

Mayr, E. and Ashlock, P.D. (1991). *Principles of Systematic Zoology*, 2nd edition. McGraw-Hill, New York.

McKeachie, W.J., Lin, Y.G., Strayer, J. (2002). Creationist vs. evolutionary beliefs: Effect on learning biology. *American Biology Teacher*, 64, 189–192.

Page, S.L. and Goodman, M. (2001). Catarrhine phylogeny: Noncoding DNA evidence for a diphyletic origin of the mangabeys and for a human-chimpanzee clade. *Molecular Phylogenetics and Evolution*, 18(1), 14–25.

Perrenoud, P. (1993). Curriculum : le formel, le réel, le caché. *La pédagogie : une encyclopédie pour aujourd'hui*, 61–76.

Pobiner, B. (2016). Accepting, understanding, teaching, and learning (human) evolution: Obstacles and opportunities. *American Journal of Physical Anthropology*, 159, 232–274.

Quessada, M.P. (2008). L'enseignement des origines d'Homo sapiens, hier et aujourd'hui, en France et ailleurs : programmes, manuels scolaires, conceptions des enseignants. Doctoral Dissertation, Université Montpellier II-Sciences et Techniques du Languedoc.

Quessada, M.P. and Clément, P. (2007). An epistemological approach to French syllabi on human origins during the 19th and 20th centuries. *Science & Education*, 16(9), 991–1006.

Reiss, M.J. (2008). Should science educators deal with the science/religion issue? *Studies in Science Education*, 44, 157–186.

Reiss, M.J. (2011). How should creationism and intelligent design be dealt with in the classroom? *Journal of Philosophy of Education*, 45, 399–415.

Rey, A., Minier, L., Malassis, R., Bogeart, L., Fagot, J. (2018). Regularity extraction across species: Associative learning mechanisms shared by humans and non-human primates. *Topics in Cognitive Science*, 11(3), 573–586.

Ross, A. (2000). *Curriculum: Construction and Critique*. Falmer Press, London.

Sanders, M. and Makotsa, D. (2016). The possible influence of curriculum statements and textbooks on misconceptions: The case of evolution. *Education as Change*, 20(1), 1–23.

Shoshani, J., Groves, C.P., Simons, E.L., Gunnell, G.F. (1996). Primate phylogeny: Morphological vs. molecular results. *Molecular Phylogenetics and Evolution*, 5(1), 102–154.

Silva, H.M., Penaloza, G., Tomasco, H., Carvalho, G.S. (2018). Chimpanzee included in the genus *Homo*. How biology teachers from three Latin American countries conceive it. *Journal of Biological Education*, 53(5), 506–515.

Simpson, G.G. (1963). The meaning of taxonomic statements. In *Classification and Human Evolution*, Washburn, S.L. (ed.). Aldine, Chicago, IL.

Skoog, G. and Bilica, K. (2002). The emphasis given to evolution in state science standards: A lever for change in evolution education? *Science Education*, 86(4), 445–462.

Vauclair, J. and Cochet, H. (2013). Ontogeny and phylogeny of communicative gestures, speech-gestures relationships and left hemisphere specialization for language. In *Oxford Studies in the Evolution of Language: The Evolutionary Emergence of Human Language*, Botha, R. and Everaert, M. (eds). Oxford University Press, Oxford.

Wagner-Egger, P., Delouvée, S., Gauvrit, N., Dieguez, S. (2018). Creationism and conspiracism share a common teleological bias. *Current Biology*, 28(16), 867–868.

Watson, E.E., Penny, D., Easteal, S. (2001). *Homo* genus: A review of the classification of humans and the great apes. *Homo Genus*, 1000–1012.

Whiten, A. (2017). Culture extends the scope of evolutionnary biology in the great apes. *Proceedings of the National Academy of Sciences of the United States of America*, 114, 7790–7797.

Wildman, D.E., Uddin, M., Liu, G., Grossman, L.I., Goodman, M. (2003). Implications of natural selection in shaping 99.4% nonsynonymous DNA identity between humans and chimpanzees: Enlarging genus *Homo*. *Proceedings of the National Academy of Sciences of the United States of America*, 100(12), 7181–7188.

Wiley, E.O. and Lieberman, B.S. (2011). *Phylogenetics: Theory and Practice of Phylogenetic Systematics*. John Wiley & Sons, New York.

Young, M. (2014). What is a curriculum and what can it do? *The Curriculum Journal*, 25(1), 1–7.

Zeitoun, V. (2015). Les paléoanthropologues sont-ils en meilleure position que les autres pour se permettre d'ignorer les règles de la systématique ? Un bref historique. *Biosystema*, 30.

Academic programs

BO no. 1, 22 January 2019.

BO no. 8, 25 July 2019.

BO no. 11, 26 November 2015.

BO no. 30, 30 July 2020.

6

The Current "Synthesis versus Extended Theory of Evolution" Controversy: A Training Opportunity about the Nature of Science

Magali FUCHS-GALLEZOT[1] and Corinne FORTIN[2]

[1] *Didasco, Université Paris-Saclay, France*

[2] *LDAR Université Paris-Cité, Université de Paris-Est-Créteil, France*

6.1. Introduction

For several years, Earth and Life Sciences (ELS) curricula in France have emphasized the need for students to "understand the nature of scientific knowledge and its methods of elaboration" (MEN 2019). There is thus a professional stake in training teachers to better understand and teach the epistemological dimensions of science, including how scientific knowledge is developed. Too often, science is still taught as neutral, stable knowledge, without taking into account the conflicts that have animated it or that still animate it. This results in an approach to science teaching that is mainly oriented towards the appropriation of concepts, without necessarily having a comprehensive and critical view of the way science works. In this chapter, we are interested in the Nature of Science (NoS), i.e. in a field of research that deals with the teaching and learning of the idea of science and with the understanding of the practices of production of scientific knowledge.

The Evolution of Life: Teaching, Learning and Training,
coordinated by Corinne FORTIN and Julie GOBERT. © ISTE Ltd 2023.

We explore a facet of teacher training that is still underdeveloped, that of contemporary scientific controversies. Generally focused on the acquisition of scientific results or on key episodes in the history of science, teacher education rarely takes into account contemporary controversies. At most, they are mentioned to point out the uncertainty of emerging knowledge (Dunlop and Veneu 2019). Yet, since the 2000s, a question has been causing agitation in the scientific community in evolutionary biology: should the modern evolutionary synthesis theory (MEST) be expanded into an extended evolutionary synthesis theory (EEST) or not? It is not up to us to answer this question, nor even to take part in the controversy, but we wish to seize it as an opportunity to explore its didactic potential in terms of teacher training in the field of NoS.

The choice of this controversy is also based on the fact that it involves a competition between the epistemic validities of different approaches within the scientific community. On the one hand, the aim is to explore the management of disagreements and the uncertainties that run through it, and on the other hand, the aim is to start a reflection on the didactic potential of the analysis tools proposed on this occasion for the training of ELS teachers. Despite the recognized challenges for teaching ABOUT sciences[1], science teaching is still mainly centered on a teaching *OF* sciences[2]. Based on the analysis of an article published in *Nature* (2014) entitled "*Does evolutionary theory need a rethink?*"[3], we wish to explore this controversy from the point of view of the NoS and thus contribute to opening up possibilities for thinking about teaching *ABOUT* science, in particular concerning the role of theoretical pluralism in the construction of scientific knowledge.

First, we will discuss the NoS characteristics, then the characteristics of scientific controversies, and finally, we will examine the synthetic versus extensive theory of evolution controversy from the perspective of NoS teacher training.

6.2. Contemporary controversies: between uncertainties and disagreements

Contrary to historical controversies, whose outcome is already known because they are reconstructed, retrospectively, from scientifically proven knowledge, contemporary scientific controversies, because of the instability of emerging knowledge, constitute:

1. ABOUT the sciences refers to the epistemological dimension.
2. OF the sciences refers to scientific concepts.
3. https://www.nature.com/articles/514161a.

> the great way to enter into the science that is being done. If we consider controversies not as fights, but as moments where we start by not knowing, and where we discuss, it is the royal road by which to enter into scientific activity (Latour 2006, p. 118).

They thus have an educational and formative value insofar as their teaching makes it possible to understand "science in action" (Latour 1987). The choice of using a contemporary scientific controversy as a reference allows teachers in training to be immersed in a situation where there is still uncertainty, the unknown, in order to activate an authentic intellectual tension (Yacek 2018).

Thus, as Dunlop and Veneu (2019) remind us, contextualizing a NoS teaching situation with a contemporary case helps to focus learning on the alternatives at play in the controversy, not just the provisional or enduring nature of knowledge (Allchin 2011). However, it is not without risk because teachers cannot resolve scientific uncertainties or evaluate empirical evidence. As a result, it leaves the open and uncertain cases hanging at the end of the teaching situation (Allchin et al. 2014). For Duschl (2008) and Sandoval (2014), controversy is, nevertheless, an example of an investigative practice of scientific activity where knowledge not only results from experimental arbitration, but also obeys the criteria of confrontation of observational and experimental facts, ideas and exchanges within a community in order to control constructed scientific knowledge. The controversy, as a scientific activity, is therefore part of a given society, at a given time, and contributes to not freezing knowledge in order to make it evolve, either by continuity or by rupture (Khun 1970). In a study of social representations of scientific controversies conducted with teachers, Dunlop and Veneu (2019) show that there is a strong contrast between the perceived importance of controversies in scientific practices and the secondary character they attribute to them in their teaching. Teachers perceive controversies as unresolved "live" problems due to a lack of procedure or as emerging from "peer review" work. The lack of data and evidence is also mentioned. Some teachers associate controversies with outdated and ridiculous historical ideas rather than explanatory systems relevant to their time. In particular, a barrier to teaching controversies was identified:

> The discourse around "teaching the facts" conveys an *image of science* that is both singular and static, and of *knowledge creation* in science as incremental and constructive (Dunlop and Veneu 2019, p. 701).

However, there is also the need to support a certain confidence in science and thus to provide a more "certain" image of science. Furthermore, they identified various difficulties in dealing with scientific controversies in the discourse of the teachers interviewed and opened up avenues such as presenting the different positions in order to allow students to form their own opinions. Oulton et al. (2004), however, point out that this attempt to be balanced is often difficult and that it is probably more relevant to develop a critical attention to bias. The issue may be more about capturing the modalities of controversies within the community and the role of scientists. Dunlop and Veneu (2019) also point out that it might be more relevant to focus more on the controversial claims than on the controversial problem, to focus instead on the modalities of scientific knowledge elaboration, the role of evidence, the structure of arguments, where the controversy is played out, between whom and on what basis, without forgetting to take into account the social context.

At the end of their research, these authors propose, in particular, to pay attention to selecting a controversy that is adapted to the target audience, particularly from a scientific point of view, as well as to clarify what the controversy is about, where the disagreement lies and why. Working on the controversy would thus make it possible to focus on the way it is played out within the scientific community, between which actors, on what basis(bases) and with what argumentation. We consider with Dunlop and Veneu (2019) that scientific controversies are essential aspects of scientific practices integral to the NoS, and that teaching them requires teachers not only to have knowledge, but also to use questioning and argumentation to evaluate the assertion of the knowledge at stake.

6.3. Nature of science through contemporary controversies

6.3.1. *Teachers' visions of science and difficulties in teaching NoS*

NoS research shows that the visions of pupils, students and teachers most often do not conform to contemporary conceptions of scientific activity and are very often heterogeneous. Roletto thus described the heterogeneity of the views of future French teachers concerning the idea of science:

> The majority of the subjects share an empiricist point of view, as to the nature of scientific knowledge; a realist one, as to its objects; an inductivist one, as to its approaches; a relativist and contextualist one, as to its status. Moreover, the subjects give very little importance to the social dimension of science and to the creative aspects of the scientists' activity (Roletto 1998, p. 11).

Various NoS research studies show that several factors influence the "transposition" of teachers' knowledge and skills on science into their practice.

These include factors specific to the field, such as a lack of resources and experience in understanding NoS, and also more general factors such as a lack of time to implement the curriculum, classroom management and organizational constraints, concerns about student motivation and abilities, teaching experience, institutional constraints (Abd-El-Kalick and Lederman 2000) and a lack of knowledge of the history and philosophy of science (Collins et al. 2001). In other words, while improving teachers' understanding of scientific practices, of how scientific knowledge is developed and of the nature of scientific knowledge is necessary, it does not seem to be sufficient to promote teaching *ABOUT* science.

It is therefore in the perspective of a teacher training aimed at both the acquisition of knowledge and skills ABOUT NoS, and at opening up teaching possibilities that our contribution fits.

6.3.2. *Characterization of NoS for thinking about training content on science*

While there is unanimity in the educational world, at the international level, about the need for teaching about NoS, there are many debates in the NoS field about the choices of a characterization of science and what to teach *ABOUT* science in order to develop and appropriate a scientific culture (Deng 2011).

In the field of NoS, Lederman (2007) refers to the epistemology of science, to science as a way of knowing and to the values and claims inherent in scientific knowledge and its developments (Lederman and Zeider 1987). This approach is based on key aspects[4]: accessibility to students, their recognition by the public and their usefulness for citizens.

This approach is very fruitful in terms of research, but it is nevertheless rivaled by others that propose conceptualizations more focused on the characteristics of the sciences. Three other approaches have thus been proposed:

– "Social practices" (Maurines et al. 2013) make explicit the different dimensions of scientific practices that contribute to the elaboration of scientific knowledge by taking into account both the people involved and the history of these practices.

---

4. Tentativeness of Scientific Knowledge, Observations and Inferences, Subjectivity and Objectivity in Science, Creativity and Rationality in Science, Social and Cultural Embeddedness in Science, Scientific Theories and Laws, Scientific Methods (Lederman 2007).

– The “Whole science” approach (Allchin 2011) identifies a list of items that account for the set of dimensions considered at the individual, collective and institutional levels to assess the reliability of scientific claims. Allchin’s approach thus proposes a “functional scientific literacy” that is not limited to the acquisition of declarative knowledge about science (Lederman 2007).

– The “Family Resemblance Approach” (Erduran and Dahger 2014; Irzik and Nola 2014), which characterizes NoS through different categories focused on epistemic, institutional and social aspects of science.

In the perspective of our contribution, which is aimed at initiating a reflection on teacher training that will allow readers to develop their understanding of the EEST/MEST controversy, their ability to analyze it in its various dimensions and to open up possibilities for their practice, it seemed relevant to us to retain:

– On the one hand, the “social practices” approach, which provides a grid for analyzing the various dimensions of the controversy and thus proposes broad categories that support a multidimensional understanding of the controversy (Table 6.1). The choice of the “social practices” approach rather than the “Family Resemblance Approach” is aimed at understanding scientific controversies as scientific practices.

Dimension of social practices	Characterization
Epistemological dimensions	ACR: Aims (A), general characteristics of knowledge (presuppositions and values) (C), relations between science and other fields (R)
	Objects of study and problems (O-Pb)
	Products obtained (P)
	Resources mobilized (Re)
	Elaboration (E): modalities of construction and validation of knowledge (activities, approaches, rules)
Sociological dimensions	Scientific community (Co)
	Society (S)
Psychological dimensions	Attitudes and qualities of the actors (A)
Historical dimensions	History (H)

Table 6.1. *Science as practices of a community of people acting as scientists (from Maurines et al. 2013)*

– On the other hand, there is the "whole science" approach, which provides a more precise analysis grid focused on evaluating the reliability of scientific claims, an aspect that seems to us to be central in the case of a controversy. Here, we summarize the main points of the "whole science" grid with examples of items relating to the different dimensions of reliability in science, from the individual to the institution (Table 6.2).

Dimensions of reliability in Science	Examples of items
Observation and reasoning	Completeness of evidence; robustness (agreement among different types of data); alternative explanations, etc.
Methods of investigation	Controlled experiment; replication and sample size; correlation/causation; etc.
History and creativity	Role of imagination and creative synthesis Role of analogy, interdisciplinary thinking; error and uncertainty, conceptual shifts; consilience with established evidence; etc.
The human context	Motivation; personality; etc.
Culture	Cultural beliefs: ideology, religion, nationality, role of gender and class bias, etc.
Social interactions among scientists	Collaboration; competition; peer review; critique; resolving disagreements; form of persuasion; limits of alternative theoretical perspectives, academic freedom, etc.
Cognitive processes	Confirmation bias; role of prior beliefs; emotional versus evidence-based perceptions of risk; etc.
Economics/fundings	Sources of research funding; conflict of interest; etc.
Instrumentation and experimental practices	New instruments and their validation; ethics of experimentation; etc.
Communication and transmission of knowledge	Credibility of scientific journals and media; social responsibility of scientists; etc.

Table 6.2. *Whole science grid (from Allchin 2011)*

These two approaches, although one belongs to Social Practices and the other to Reliability in Science, seem to us to be able to be put into perspective insofar as they both apprehend, in a complementary way, the sciences at different levels of their practice (from the individual to the institution), while also including their historical dimension. Table 6.3 proposes such a perspective by placing the individual at the center of the practices.

Maurines et al. (2013) Dimensions of social practices		From Allchin 2011 Dimensions of reliability in science
Epistemological dimensions of scientific practices (aims and values, objects, intellectual and instrumental resources, products, development)		Observation and reasoning
		Methods of investigation
		Instrumentation and experimental practices
Sociological dimensions of scientific practices (community and society)		Communication and **transmission of** knowledge
		Social interactions among scientists
		Economics/funding
		Culture
Psychological dimensions of the actors of the practices (attitudes, qualities)		Cognitive processes
		The human context
Historical dimensions of practices (history)		History and creativity

Table 6.3. *Putting the NoS approach into perspective through Maurines et al.'s (2013) "Social Practices" and Allchin's (2011) "Whole Science"*

In the context of the MEST/EEST controversy, these two grids should allow a complementary analysis of this controversy from the point of view of NoS.

6.3.3. *Characterization of contemporary scientific controversies*

Before discussing the MEST/EEST controversy, we would like to return to the characterization of the scientific controversy. Without excluding the historical and sociocultural context of controversies, our resolutely didactic approach is focused on the knowledge at stake that divides the scientific community, rather than on a sociological approach mainly centered on the interplay of actors and institutions (Chateauraynaud 2011; Raynaud 2018). In the case of the MEST/EEST controversy that interests us here, it cannot be understood as a disagreement between personalities of the scientific community, even though otherwise prominent personalities embody choices in favor of or against the extension of MEST. We understand the controversy in the sense proposed by Dascal (1998, 2008), i.e. as an epistemological competition between two research programs, in this case between MEST and EEST.

Indeed, if every controversy is partly, but only partly, based on a disagreement due to uncertainties linked to a lack of knowledge, data or solutions; not every disagreement necessarily leads to a controversy. To situate the controversy as an object of didactic study, we retain, here, the distinctions proposed by Dascal:

> – Controversy is not a "dispute" that would end as a result of arbitration, for example, in a dispute over the priority of a discovery.
>
> – Nor is it an argued "discussion" centered on a possible "error" in problem solving involving calculations, experiments or data. In this case, the discussion is decided by a probative demonstration, the contribution of new data or a decisive experimentation which corrects or not the presumed error.
>
> – It involves a profound divergence in existing methods of resolution associated with new concepts. Contrary to the dispute or the discussion, it cannot be decided only by empirical solutions or arbitration, because it calls into question the coherence of an existing explanatory framework by creating alternative research programs. Controversy then becomes "the place where critical activity takes place, essential for the rational progress of knowledge", (Dascal 1998, p. 103)

and a means to better understand what is problematic.

In this sense, contemporary scientific controversy is based on a persistent conflict, in the scientific community, to defend a research program (Donato Rodríguez and Bonilla 2014). It is also a dialogic and argumentative activity with the aim of elaborating or revising achievements through new heuristics (Dunlop and Veneu 2019). It is then translated by a setting in tension that can lead to a partial or total rupture with a dominant explanatory scheme. Controversy thus appears to be a practice of confronting competing explanations to meet a need for intelligibility. As Dascal (2008) points out, it is not a question of artificially creating categories that would have a demarcation value in relation to one another, since intersections are always possible between dispute, discussion and controversy. It is mainly a question of having indicators to identify the dominant and the emerging form.

6.4. Main epistemological issues of the MEST/EEST controversy

The MEST introduced in the 1930s and 1940s combines random mutations, natural selection and population genetics (Fisher 1930; Wright 1931; Haldane 1932; Dobshanky 1937; Muller 1940) to account for adaptation and the evolutionary history of life. With paleontology (Simpson 1944), it takes into account evolutionary rhythms on a geological scale, both slow and gradual, through a cumulative effect of microevolution into macroevolution. It also profoundly renewed the concept of species by definitively breaking with essentialism in favor of the biological species (Mayr 1942) and revisited in depth the modes of speciation (Mayr 1982).

In the 1970s and 1980s, the neutralist theory of Kimura (1983) shook up the unifying explanation of natural selection by showing that some mutations could be selectively neutral and that the genetic transformations of populations could also be explained by random drift. MEST then integrated genetic drift into its explanatory scheme while keeping the accumulation of mutations under the action of natural selection as the main explanation of evolution.

The work of Eldredge and Gould (1972) mobilized the mutations of developmental genes to challenge the gradualist view of MEST according to which macroevolution would only be the result of the accumulation of microevolution. A new field of research then emerged, that of evolutionary developmental biology or "evo-devo" (Gould 1977; Sultan 2007). It gives development a determining role in the evolutionary history of life. Macro-evolutionary changes were not only the result of the accumulation of selected mutations, but also of interactions between the environment and the development of organisms. This leads, according to these authors, to the envisaging of extremely rapid macroevolutionary changes independently of the cumulative effect of microevolution. As Gilbert (2000) points out, in this case, it is no longer a question of the survival of the fittest, but of the possible emergence of the fittest, through the innovations produced during development. The pre-eminence of frequency changes in populations under the effect of natural selection or drift is thus competed for by the selection of phenotypes resulting from developmental modifications. A new stage is opening up in the history of evolutionary biology insofar as selection is no longer exercised only on genes, but also on phenotypes.

In the 2000s, the work of West-Eberhard (2003, 2005) revisited Waddington's 1940s work on epigenetics and phenotypic plasticity and re-examined the role of the environment in evolutionary changes. Indeed, in the case of phenotypic plasticity, the same genotype can express different phenotypes in response to environmental variations. Beyond accommodation, in the sense of an adaptive phenotypic adjustment in response to environmental variations, some phenotypes have a higher fitness than other plastic phenotypes. This adaptive plasticity can then be subject to phenotypic selection. This is what West-Eberhard (2005) says about phenotypic plasticity when she concludes that genes are probably followers rather than leaders in evolution[5]. In this case, it is no longer the mutations that are selected, but rather an alternative phenotype resulting from the genetic expression of the same genotype. There is thus a developmental bias, i.e. phenotypic variants appear more easily than others depending on environmental variations, and the selected phenotype can be genetically fixed and transmitted to subsequent generations. The evolutionary impacts of interactions between development and ecological adaptations have thereby opened up a new field of investigation for evolutionary theory called "Eco-Devo-Evo" (Sultan 2007, 2015). For this reason, Pigliucci (2007) considers that MEST no longer meets the new research challenges, and needs to be expanded into EEST. Following West-Eberhard's work, Lalland et al. (2015) has led not only to a thorough revision of MEST, but also to an explanatory alternative that challenges the foundations of neo-Darwinism.

In summary, proponents of MEST and EEST argue for two explanatory approaches to evolution: one primarily gene-centric and the other more phenocentric. Consider, for example, the cichlidae of Lakes Victoria, Malawi and Tanganyika, which illustrate cases of rapid speciation over the last 200,000 years (Meyer et al. 1990). Phenotypically similar cichilidae species in the three lakes occupy similar niches, but do not share a direct common ancestor. Two interpretations are then possible:

– That of a convergent evolution according to the MEST explanatory scheme: the common phenotypic traits in the three lakes are the result of the action of selection in similar ecological niches. This leads to an independent evolution of similar phenotypic traits, with independent genetic bases, where genetic mutations would be selected over time and would have produced the same phenotypic effects in the three lakes. In this case, trait diversification would result from mutational selection.

– An evolution linked to phenotypic plasticity according to the EEST explanatory scheme: an ancestor with phenotypically plastic morphological traits could be at the origin of the current variability. Variation in plastic phenotypic traits in relation to ecological niche diversity would then have led to the selection of plastic phenotypic traits and to phenotypic diversity in the three lakes. In this case,

5. "Thus, genes are probably more often followers than leaders in evolutionary change."

trait diversification would result from the selection of plastic phenotypes, genetically fixed by genetic accommodation.

For the proponents of MEST, evolution is described in terms of genetic variations driven by natural selection or a drift in populations. This neo-Darwinian explanatory model is based on selection that acts at the level of genes through microevolution. MEST is part of the genetic heredity framework where the development of organisms is a product of evolution. Phenotypic plasticity and niche construction are then understood as consequences, not possible causes, of evolution (Table 6.4).

For the proponents of EEST, selection would not only act on mutations (microevolution), but also on alternative phenotypes resulting from developmental processes in interaction with environmental variations. The de facto consequence of this is the decoupling of microevolution from macroevolution, which is the core of the neo-Darwinian model, since mutation selection would no longer be the only one involved.

	EEST	MEST
Aim	Going beyond the gene-centric vision to include a phenocentric vision	Maintaining the gene-centric view
Concepts	Development bias Plasticity Niche construction Epigenetics	Mutation Natural selection Genetic drift Gene flow
Environmental role	The interactions of organisms with the environment are part of the evolutionary process	The environment can modify selection, but is not part of the evolutionary process
Causes/results of evolution	Developmental biases, plasticity or niche construction, epigenetics, are evolutionary causes	Developmental biases, plasticity or niche construction, epigenetics, are the results of evolution
Tension within the community	Need to extend the current theory to include multilevel selection (plasticity, epigenetics, etc.)	Maintaining neo-Darwinism with only a small expansion to incorporate plasticity, epigenetics, etc. as evolutionary background noise

Table 6.4. *The main dissensus between synthetic and extensive theory of evolution*

According to EEST, evolvability, i.e. the capacity to produce heritable phenotypic variations, would not be limited to mutation or genetic drift. It would also integrate developmental plasticity and its phenotypic effects, as well as non-genetic heredity with epigenetics to account for the production of phenotypic variations and their selection. In this case, natural selection is no longer the dominant explanation of evolution, since development is also under the control of interactions between organisms and their environment. As a result, the distinction made by Mayr (1961, 1993) between proximal and distal causes at the heart of MEST is at a minimum revisited (Otsuka 2015; Scholl and Pigliucci 2015) and at a maximum abandoned in favor of a so-called reciprocal causality (Laland et al. 2011; Mesoudi et al. 2013). Indeed, in the EEST framework, proximal causes (for example, phenotypic changes) are induced under environmental variation, and distal causes (for example, selection for plastic phenotypes) are not the products of natural selection, mutations or genetic drift. However, they are possibly the cause of new evolutionary directions through the action that the environment exerts on their development and by the action that they exert on their environment to transform it.

The MEST/EEST confrontation is therefore indeed a scientific controversy in the sense given by Dascal (1998, 2008) because even though there is no abandonment of neo-Darwinism, there is a profound paradigmatic divergence. What is at stake, here, is a reconfiguration of the current theory of evolution in the sense that some refer to EEST as "post-synthesis" (Jablonka 2005, 2006) or "beyond neo-Darwinism" (Noble 2015). According to Noble, the epistemological tension between MEST and EEST rests on the gene-centric view of MEST that would, in his view, prevent an integration with EEST, by downplaying the role of epigenetic inheritance, multilevel selection and genomics in evolutionary processes. This would lead to a breaking point in terms of the limits of extension for MEST, hence the subject of the controversy that is now driving the evolutionary biology community.

The MEST/EEST controversy thus allows us to move beyond the "monstration" of evolutionary history that is still very much embedded in the field of evolutionary education and teaching (Fortin 2015).

6.5. The MEST/EEST controversy: an opportunity for teacher education?

In the case of the MEST/EEST controversy, confrontation between competing explanations is an integral part of the social practice of the scientific community (Martinand 1986). This confrontation manifests itself during colloquia or in academic journals (Wazeck 2013) and is structured into a debate that becomes a collaborative space where the protagonists put forward their arguments. This practice of knowledge construction, far from being optional, is intrinsically linked to the construction of knowledge, and as such appears to be an integral component of the NoS.

From a teacher training perspective, the NoS framework is also aimed at problematizing theoretical pluralism by fostering authentic critical activity. We also draw on an article entitled "Does evolutionary theory need a rethink?" published in 2014, in the journal *Nature*, to explore its potential for NoS teachers' training.

This controversy appears to be a training opportunity aimed, in a context of uncertainty, at allowing teachers to better identify the issues at stake in the controversy, and also to better understand how scientific controversies unfold in scientific practices, in order to gain operationality in their analysis.

The double set of grids associating the "social practices" (Maurines et al. 2013) and the "whole science" approaches (Allchin 2011) made available to teacher education should make it possible to reach these objectives. As these grids are generic, they can potentially be reused to develop teaching ABOUT science, either to formulate objectives based on the diversity of the categories and items proposed or by remobilizing them after adaptation.

The article published in the "How" section of *Nature* gives the floor to representatives of the two scientific communities involved in the MEST/EEST controversy. It is the product and the reflection of the socio-scientific practices. The aim is to present the different points of view, in particular with regard to theoretical divergences, and also to the practices involved. Strictly speaking, this is not a dialogue or a debate between the two parties, but rather a face-to-face meeting of the two points of view. Each of the two parties independently puts forward its arguments, the better to defend the necessity or not of an extension of the current theory of evolution.

Tables 6.5 to 6.7 present examples of results obtained from the analysis of the *Nature* article, conducted using the double set of analysis grids (Table 6.3) of Allchin (2011) and Maurines et al. (2013). For reasons of space, we cannot present the full analysis of the article here. Each of the tables reports on one aspect of the analysis, and for each of these aspects, some examples of the elements identified in the article are mentioned.

Table 6.5 reports on the analysis of the form of the article, the context in which the protagonists express themselves, and the different dimensions in terms of "social practices" and "whole science": what form did the controversy take within the community? Which journal published this article? How does its form reflect the controversy that runs through the scientific community? What are the types of arguments used by the protagonists of the controversy?

<table>
<tr><th>Social practices Maurines et al. (2013)</th><th>EEST</th><th>MEST</th><th>Whole science, (Allchin 2011)</th></tr>
<tr><td rowspan="3"><u>Sociological dimensions</u>

Community</td><td colspan="2">Status of the journal Nature</td><td><u>Communication</u>
Credibility of scientific journals</td></tr>
<tr><td colspan="2">"How to" section
Title: formulated as a question: "Does evolutionary theory need a rethink?"
Subtitle: makes explicit what the controversy is about: "researchers are divided over what processes should be considered fundamental"
Presentation of the arguments in parallel of the two points of view in debate
Point: Yes, Urgently Counterpoint: No, all is well
Signature of the article: by groups of scientists led by one or two emblematic figures for whom are indicated title, laboratory, institution, country (e.g. "Kevin Laland is professor of behavioural and evolutionary biology at the University of St Andrews, UK", Wray: professor of biology at Duke University in Durham, North Carolina, USA)</td><td><u>Social interactions among scientists:</u>

Collaboration; competition; resolving disagreements</td></tr>
<tr><td>Reference to Darwin as a common ground with MEST to better show the proposed theoretical renewal.</td><td>Reference to Darwin throughout the text
=> authority figure: "like Darwin, we think" => to show the inclusive nature of evolutionary MEST</td><td>Form of persuasion</td></tr>
</table>

Table 6.5. *Contextual analysis grid for the article*

Table 6.6 shows the analysis of the diversity of the dimensions of the scientific practices mobilized among those we identified as mentioned by the protagonists of the controversy.

Social practices Maurines et al. (2013)	EEST	MEST	Whole science, Allchin (2011)
Epistemological dimensions Elaboration	"The data supporting our position gets stronger every day" (p. 162)	"Much less clear is whether plasticity can 'lead' genetic variation during adaptation" (p. 163)	Observation and reasoning
Intellectual, instrumental resources, etc.	"Strong support comes from […] developmental biology, but also genomics, epigenetics, ecology and social science" (p. 162)	"Evolutionary biologists today draw inspiration from fields as diverse as genomics, medicine, ecology, artificial intelligence and robotics" (p. 163)	History and creativity Interdisciplinary thinking
Sociological dimensions Community	"In the time since, as members of an interdisciplinary team, we have worked intensively" (p. 161)	"Most evolutionary biologists have a list of topics that they would like to see given more attention" (p. 163)	Social interactions among scientists academic freedom
Psychological dimensions Attitude	Hostility of some evolutionary biologists	Sarcastic attitude (earthworms) Or a call to humility?	Human context
Historical dimension		Reference to Darwin and the history of evolutionary research	

Table 6.6. *Examples of results obtained from the analysis of the controversy showing the diversity of the dimensions of scientific practices mobilized/mentioned by the protagonists of the controversy*

Table 6.7 shows results of the dual analysis of the epistemological issues of the controversy.

Social practices Maurines et al.	EEST	MEST	Whole science, Allchin
Epistemological dimensions	Division on the processes to be considered fundamental to explaining evolution?		
Theoretical elaboration/ Product	"Without an extended evolutionary framework, the theory neglects key processes" (p. 161) "An alternative vision of evolution is beginning to crystallize, in which the processes by which organisms grow and develop are recognized as causes of evolution" (p. 161) => **Geno-centric vision override**	"Theory accommodates evidence through relentless synthesis" (p. 161) Furthermore, the phenomena that interest Laland and colleagues are just four among many that offer promise for future advances in evolutionary biology (p. 163) => **Maintain geno-centric vision**	Social interactions among scientists limits of theoretical alternatives
Characteristic aims Predictive power, explanatory power of a theory	"For example, cichlid fishes in L. Malawi are more closely related to other cichlids in L. Malawi than to those in L. Tanganyika, but species in both lakes have strikingly similar body shapes" (p. 162) "SET explains such parallels as convergent evolution: similar environmental conditions select for random genetic variation with equivalent results. This account requires extraordinary coincidence to explain the multiple parallel forms that evolved independently in each lake. A more succinct hypothesis is that developmental bias and natural selection work together. Rather than selection being free to traverse across any physical possibility, it is guided along specific routes opened up by the processes of development" (p. 162)	"So, none of the phenomena championed by Laland and colleagues are neglected in evolutionary biology. Like all ideas, however, they need to prove their value in the marketplace of rigorous theory, empirical results and critical discussion. The prominence that these four phenomena command in the discourse of contemporary evolutionary theory reflects their proven explanatory power, not a lack of attention" (p. 163)	Observation and reasoning alternatives explanations

Table 6.7. *Examples of results obtained from the analysis of the article showing epistemological issues of the controversy*

The analysis of this *Nature* article allows teachers not only to access the epistemological context of the controversy, but also to understand the nature of the competition between EEST and MEST, in particular, the distinction between expansion claimed by proponents of MEST, and the extension defended by proponents of EEST. As Pievani (2012) points out, what is problematic is that EEST appears to be an alternative research program to MEST, not an increase and theoretical readjustment as was the case for the transition from Darwinism to neo-Darwinism, hence the focus of the controversy.

It is therefore the weighing up of the arguments of a part of the community that will tip the balance (temporarily or definitively) in favor of MEST or EEST, or will maintain a form of status quo with the cohabitation of the two theoretical frameworks. However, it is also the dynamics of the functioning of the scientific community, in these modes of production and dissemination of knowledge, between competition and collaboration, that will determine the outcome of the controversy.

The analysis of this contemporary controversy, far from being resolved, is an opportunity to allow trainee teachers to become aware that uncertainty is inseparable from the ways in which scientific knowledge is developed. It is also an opportunity to train them to take into account the different aspects/dimensions of scientific practices that support the robustness of the knowledge at stake, in particular the links between theoretical elaboration and socio-scientific interactions.

6.6. Conclusion

In a period of questioning of scientific knowledge, of scientists themselves and of science, training teachers about the different dimensions of the NoS becomes a priority. On the one hand, because the understanding of the theoretical issues of the controversy could facilitate a possible update of scientific knowledge, and on the other hand, because a better understanding of the epistemological aspects could also support scientific learning.

The choice of the contemporary MEST/EEST controversy as a training medium for teachers can thus contribute to a better appropriation of the scientific field studied, and also and above all allow them to better understand the processes of production of scientific knowledge at stake, in particular through the identification of the different dimensions of NoS, in order to then be able to really take charge of them in their teaching.

6.7. References

Abd-El-Khalick, F. and Lederman, N.G. (2000). Improving science teachers' conceptions of nature of science: A critical review of the literature. *International Journal of Science Education*, 22(7), 665–701.

Allchin, D. (2011). Evaluating knowledge of the nature of (whole) science. *Science Education*, 95(3), 518–542.

Allchin, D., Andersen, H.M., Nielsen, K. (2014). Teaching nature of science: Integrating student inquiry, historical cases, and contemporary cases in classroom practice. *Science Education*, 98(3), 461–486

Chateauraynaud, F. (2011). Sociologie argumentative et dynamique des controverses : l'exemple de l'argument climatique dans la relance de l'énergie nucléaire en Europe. *A Contrario*, 2, 131–150.

Collins, S., Osborne, J., Ratcliffe, M., Millar, R., Duschl, R. (2001). What "ideas-about-science" should be taught in school science? A Delphi study of the expert community. *Annual Conference of the American Educational Research Association*, Seattle, 10–14 April.

Dascal, M. (1998). The study of controversies and the theory and history of science. *Science in Context*, 11(2), 147–154.

Dascal, M. (2008). Dichotomies and types of debate. *Controversy and Confrontation: Relating Controversy Analysis with Argumentation Theory*, 27–49.

Deng, F., Chen, D.-T., Tsai, C.-C., Chai, C.S. (2011). Students' views of the nature of science: A critical review of research. *Science Education*, 95(6), 961–999.

Dobzhansky, T. (1937). *Genetics and the Origin of Species*, 1st edition. Columbia University Press, New York.

Donato Rodríguez, X. and Bonilla, J.Z. (2014). Scientific controversies and the ethics of arguing and belief in the face of rational disagreement. *Argumentation*, 28(1), 39–65.

Dunlop, L. and Veneu, F. (2019). Controversies in science. *Science & Education*, 28(6), 689–710.

Duschl, R. (2008). Science education in three-part harmony: Balancing conceptual, epistemic and social learning goals. *Review of Research in Education*, 32(1), 268–291.

Eldredge, N., and Gould, S.J. (1972). Punctuated equilibria: An alternative to phyletic gradualism. In *Models in Paleobiology*, Schopf, T.J.M. (ed.). Freeman Cooper & Co. San Francisco, CA.

Erduran, S. and Dagher, Z. (2014). *Reconceptualizing the Nature of Science for Science Education: Scientific Knowledge, Practices and Other Family Categories*. Springer, Dordrecht.

Fisher, R.A. (1930). *The Genetical Theory of Natural Selection*, 1st edition. Oxford University Press, Oxford.

Fortin, C. (2015). Evolutionary theory in secondary schools: Some teaching issues. In *Handbook of Evolutionary Thinking in the Sciences*, Heams, T., Huneman, P., Lecointre, G., Silberstein, M. (eds). Springer, Dordrecht.

Gilbert, S.F. (2000). Diachronic biology meets evo-devo: CH Waddington's approach to evolutionary developmental biology. *American Zoologist*, 40(5), 729–737.

Gould, S.J. (1977). *Ontogeny and Phylogeny*. Belknap Press, Cambridge, MA.

Haldane, J.B.S. (1932). *The Causes of Evolution*. Harper, New York.

Irzik, G. and Nola, R. (2014). New directions for nature of science research. In *International Handbook of Research in History, Philosophy and Science Teaching*, Matthews, M. (ed.). Springer, Dordrecht.

Jablonka, E. and Lamb, M.J. (2005). *Evolution in Four Dimensions: Genetic, Epigenetic, Behavioral, and Symbolic Variation in the History of Life*. MIT Press, Cambridge, MA.

Khun, T. (1970). *The Structure of Scientific Revolutions*. University of Chicago Press, IL.

Kimura, M. (1983). *The Neutral Theory of Molecular Evolution*. Cambridge University Press, Cambridge.

Laland, K.N., Odling-Smee, J., Gilbert, S.F. (2008). Evo-devo and niche construction: Building bridges. *Journal of Experimental Zoology, Part B: Molecular and Developmental Evolution*, 310(7), 549–566.

Laland, K.N., Sterelny, K., Odling-Smee, J., Hoppitt, W., Uller, T. (2011). Cause and effect in biology revisited: Is Mayr's proximate-ultimate dichotomy still useful? *Science*, 334, 1512–1516

Laland, K.N., Uller, T., Feldman, M., Sterelny, K., Müller, G.B., Moczek, A., Strassmann, J.E. (2014). Does evolutionary theory need a rethink? *Nature News*, 514(7521), 161.

Laland, K.N., Uller, T., Feldman, M.W., Sterelny, K., Müllller G.B., Moczek, A., Jablonka, E., Odling-Smee, J. (2015). The extended evolutionary synthesis: Its structure, assumptions and predictions. *Proceedings of the Royal Society B: Biological Sciences*, 282, 20151019.

Latour, B. (1987). *Science in Action: How to Follow Scientists and Engineers through Society*. Harvard University Press, Cambridge, MA.

Latour, B. (2006). Entretien avec Bruno Latour, Fossier et Gardella. *Tracés. Revue de Sciences humaines*, 10, 113–129.

Lederman, J.S. (2007). Development of a valid and reliable protocol for the assessment of early childhood students' conceptions of nature of science and scientific inquiry. *Saarmste Executive*, 446.

Lederman, N.G. and Zeidler, D.L. (1987). Science teachers' conceptions of the nature of science: Do they really influence teaching behavior? *Science Education*, 71(5), 721–734

Martinand, J.-L. (1986). *Connaître et transformer la matière*. Peter Lang, Bern.

Maurines, L., Gallezot, M., Ramage, M.-J., Beaufils, D. (2013). La nature des sciences dans les programmes de seconde de physique-chimie et de sciences de la vie et de la Terre. *Recherches en didactique des sciences et des technologies*, 7, 19–52.

Mayr, E. (1942). *Systematics and the Origin of Species*. Columbia University Press, New York.

Mayr, E. (1961). Cause and effect in biology. *Science*, 131, 1501–1506.

Mayr, E. (1982). *The Growth of Biological Thought: Diversity, Evolution, Inheritance*. Harvard University Press, Cambridge, MA.

Mayr, E. (1993). Proximate and ultimate causation. *Biology and Philosophy*, 8, 95–98.

MEN (2019). Programme d'enseignement scientifique de la classe de première de la voie générale. *Bulletin officiel du ministère de l'Éducation nationale*, special issue 1, 22 January.

Mesoudi, A., Blanchet, S., Charmantier, A., Danchin, E., Fogarty, L., Jablonka, E., Pujol, B. (2013). Is non-genetic inheritance just a proximate mechanism? A corroboration of the extended evolutionary synthesis. *Biological Theory*, 7(3), 189–195.

Muller, H.J. (1940). Bearings of the drosophila work on systematics. In *The New Systematics*, Huxley, J.S. (ed.), Systematics Association, London.

Noble, D. (2015). Evolution beyond neo-Darwinism: A new conceptual framework. *Journal of Experimental Biology*, 218, 7–13.

Otsuka, J. (2015). Using causal models to integrate proximate and ultimate causation. *Biology & Philosophy*, 30(1), 19–37.

Oulton, C., Dillon, J., Grace, M.C. (2004). Reconceptualizing the teaching of controversial issues. *International Journal of Science Education*, 26(4), 411–423.

Pievani, T. (2012). An evolving research programme: The structure of evolutionary theory from a Lakatosian perspective. In *The Theory of Evolution and its Impact*, Faslo A. (ed.) Springer, Milan.

Pigliucci, M. (2007). Do we need an extended evolutionary synthesis? *Evolution*, 61, 2743–2749.

Raynaud, D. (2018). *Sociologie des controverses scientifiques : de la philosophie des sciences*. Éditions Matériologiques, Paris.

Roletto, E. (1998). La science et les connaissances scientifiques : points de vue de futurs enseignants. *Aster*, 26, 11–30

Sandoval, W. (2014). Science education's need for a theory of epistemological development. *Science Education*, 98(3), 383–387

Scholl, R. and Pigliucci, M. (2015). The proximate–ultimate distinction and evolutionary developmental biology: Causal irrelevance versus explanatory abstraction. *Biology & Philosophy*, 30(5), 653–670.

Simpson, G.G. (1944). *Tempo and Mode in Evolution.* Columbia University, Press, New York.

Sultan S.E. (2007). Development in context: The timely emergence of eco-devo. *Trends in Ecology & Evolution*, 22, 575–582.

Wazeck, M. (2013). Marginalization processes in science: The controversy about the theory of relativity in the 1920s. *Social Studies of Science*, 43(2), 163–190.

West-Eberhard, J. (2003). *Developmental Plasticity and Evolution.* Oxford University Press, Oxford.

West-Eberhard, J. (2005). Developmental plasticity and the origin of species differences. *Proceedings of the National Academy of Sciences*, 102(supplement 1), 6543–6549

Wright, S. (1931). Evolution in mendelian populations. *Genetics*, 16, 97–159.

Yacek, D. (2018). Thinking controversially: The psychological condition for teaching controversial issues. *Journal of Philosophy of Education*, 52(1), 71–86.

7

Historical Controversy over the Ages of the Earth and the Necessity for Deep Time for Darwinian Selection: Its Reception in Teacher Education

Patricia CRÉPIN-OBERT

LDAR Université Paris-Cité, Université de Paris-Est-Créteil, France

7.1. Context and issues

This research is focused on the development of teaching practices that give a place to the history of science in the sequences designed and implemented in the 8th grade of high school. The prescriptions for the new earth and life sciences (ELS) courses for the start of the 2016 school year[1] specify a historical dimension for all themes: planet Earth, the evolution of living organisms, the human body and health "lend themselves to the history of science, when the student situates the evolution of ideas [or knowledge] in its historical and technical context". Certain concepts are highlighted and sometimes scientists are suggested: "the shape of the

1. Special edition of Bulletin officiel No. 11 of December 24, 2015.

Earth[2], its position in relation to the sun, geological eras, continental drift (Wegener); reproduction, genetics or evolution (Darwin); vaccination and antibiotics". It is mainly an epistemological dimension of knowledge that is targeted by these lessons integrating historical elements: "Identify through the history of science and technology how scientific knowledge is constructed". Driven by these new official instructions, a continuing education course was proposed, serving as a basis for our study, the objective of which is to demonstrate that a historical resource can be a preferred support for problematizing among students. We then refocus the epistemological aims on the construction of a problem and a time of scientific controversy in order to give a certain robustness to the scientific problems treated in class.

First, we will develop how the historical construction of the problem of the age of the Earth is rich in learning perspectives for a school community, by focusing our analysis on a text by Darwin that is engaged in a controversy between naturalists and physicists.

Then, in a second step, we will clarify, from an epistemological point of view, the problematization of the age of the Earth by showing the didactic potential of the historical source.

Finally, we will specify the data collection device, within a continuing education course and based in part on this primary source. The analysis of the results obtained through the written productions and the language exchanges of the trainees will allow us to bring to light contrasting epistemological reference points via the reception of the historical source studied.

7.2. The historical construction of the problem of the age of the Earth via a founding text by Darwin

The historical source studied is a short excerpt from *On the Origin of Species, by Means of Natural Selection* (Darwin 1859, 1st English edition). Within Chapter IX, "On the Imperfection of the Geological Record", the section titled "On the vast lapse of time, as inferred from the rate of deposition and of denudation" (pp. 282–287) is of interest to us as a preferred working medium. We develop certain relevant

2. A cross-reference can be made with the requirements of the physical sciences and chemistry courses which specify the study of "Galaxies, evolution of the Universe, formation of the solar system, geological ages". Special edition of Official Bulletin No. 11 of December 24, 2015, p. 332.

elements of epistemological knowledge centered on a problem in development, intended for teachers in training.

7.2.1. *A problem federated by a scientific community*

All scientific knowledge is objectified by being subjected to peer review. In this sense, Charles Darwin (1809–1882) exposed his demonstration publicly in a naturalist context in the middle of the 19th century, to much excitement, for an emerging discipline: paleontology within geology. The choice of this founding text is due to his current presence and his notoriety at the time as a member of various institutions, a mark of scientific recognition, including the Royal Society of London[3], the Geological Society of London[4] and the Linnean Society of London[5]. Note that he was the pupil of W. Buckland (1784–1856), a famous professor of mineralogy at the University of Oxford, like C. Lyell (1797–1875) and R. Owen (1804–1892), and that they all distinguished themselves within the scientific community. The previous work and the numerous exchanges between two eminent paleontologists, W. Buckland and G. Cuvier (1769–1832), made it possible to realize the existence of very large fossil reptiles, different from crocodilians, and to assign them specifically to the strata of the Mesozoic era. In 1824, Buckland established the unpublished discovery of a very large saurian or giant lizard, which he named Megalosaurus, and published a first description in the *Transactions of the Geological Society of London*. Numerous discoveries were then made of these surprising and fantastic lizards that Owen officially named "dinosaurs" in 1842, forming a new group of gigantic extinct reptiles whose frightening size was beyond imagination, hence the term, which means "terrible lizard". In 1850, the project of a great museum of natural history in London was launched by Owen, who became its director when it opened in 1881.

At the same time, a seminal work in the community, *Principles of Geology* by Lyell, was published in 1830 and followed by numerous editions until 1872. In

3. Institution founded in 1660 for the progress and promotion of the natural sciences. Equivalent of the Academy of Sciences of Paris, founded in 1666.

4. British learned society; founded in 1807 and whose purpose is the study of the mineral structure of the Earth and the advancement of geology. It is the oldest learned society of geology in the world. Equivalent of the Geological Society of France, founded in 1830.

5. Learned society; founded in 1788 in London, dedicated to the study and dissemination of the science of classifications of living organisms or taxonomy. It is the oldest learned society dedicated to biology. In July 1858, a joint article was read by Darwin and Wallace on their respective theories of evolution, variability and the preservation of species by natural means of selection.

opposition to Cuvier's catastrophism, to explain the physical and biological changes on the surface of the globe, he defended the theory of uniform, slow and continuous real causes that must operate over an immense period of time. From the outset, in the introduction, Darwin alerts the reader to the fact that the reader must be open-minded enough to grasp geological time:

> He who can read Sir Charles Lyell's grand work on the *Principles of Geology*, which the future historian will recognise as having produced a revolution in natural science, yet does not admit how incomprehensibly vast have been the past periods of time, may at once close this volume (Darwin 1859, p. 282).

This quote announces an epistemological break in the Kuhnian sense. In the remainder of the excerpt, Darwin relies more particularly on the numerous works of eminent then-contemporary British geologists: Hugh Miller (1802–1856), James Smith (1782–1867) and Andrew Ramsay (1814–1891).

7.2.2. A problem co-constructed between facts and explanations in an evolutionary framework

The guiding thread of Darwin's demonstration is thus the construction of deep time that he knows is difficult to access through thought. His argumentative logic is based on two complementary geological phenomena: the cyclical processes of sedimentation and erosion. His reasoning is based first on the impressive thickness of the strata, which makes one aware of the depth of geological time, already announced by Buffon, and which every geologist in the field has had to grasp:

> A man must for years examine for himself great piles of superimposed strata, and watch the sea at work, grinding down old rocks and making fresh sediment, before he can hope to comprehend anything of the lapse of time, the monuments of which we see around us (Darwin 1859).

Moreover, in relation to sedimentation, Darwin accumulates clues of deep time: the thickness of the layers of conglomerates are "many thousand feet", the state of wear of the rounded gravels of which they are composed "each of which bears the stamp of time"; they are the result of the complementary phenomenon of erosion that took place in other regions, and, finally, of the sedimentary gaps there exist "enormously long blank periods" (ibid., pp. 283–284). Another example is that the decreasing granulometry prefigures a long period of erosion:

> At last the base of the cliff is undermined [by the tides], huge fragments fall down, and these remaining fixed, have to be worn away, atom by atom, until reduced in size they can be rolled about by the waves, and then are more quickly ground into pebbles, sand, or mud (ibid., p. 283).

In order to convince the reader, he uses a rhetoric that accentuates the impression of deep time.

In the end, this deep time is built to support his theory of "descent with modification", a foundation that he introduces by anticipating the major objection against it – the absence to date of intermediate fossil forms that would gradually link a chain of life:

> Independently of our not finding fossil remains of such infinitely numerous connecting links, it may be objected, that time will not have sufficed for so great an amount of organic change, all changes having been effected very slowly through natural selection (ibid., p. 282).

The mechanism of natural selection is a necessary condition that he invokes by marking a break in transformist thinking. However, this process can only operate under another condition: the deep time needed to explain the great diversity of the fossil organic forms archived in the strata and of the current living beings resulting from transformations. From an epistemological point of view, the theory of evolution that he develops puts continuity and rupture in a state of tension: to think that the continuity of species has evolved slowly over the course of time while freeing itself from an observed stratigraphic discontinuity archiving the fossils in a lacunar way:

> During each of these years, over the whole world, the land and the water has been peopled by hosts of living forms. What an infinite number of generations, which the mind cannot grasp, must have succeeded each other in the long roll of years! (ibid., p. 287).

The enigma of the age of the Earth thus stems from linked problems in stratigraphic and evolutionary paleontology – relative chronology, paleontological identity of strata, sedimentary gap, transitional form in fossils.

7.2.3. *A problem trying to be solved by quantitative reasoning*

Our choice to study the first edition[6] of *On the Origin of Species*, dating from 1859, is justified as it is the only one to propose a justified estimate of a deep time in figures: "At this rate, on the above data, the denudation of the Weald must have required 306,662,400 years; or say, three hundred million years" (Darwin 1859). His mathematical demonstration is based on two logics: the proportionality of the rule of three; and the principle of actualism promoted all his life by Lyell, a colleague and friend of his. Here Darwin relies on values of sedimentation and erosion rates given to him in correspondence with the Scottish geologist, Ramsay. The reproducibility of the estimates can therefore be subject to criticism by his readers.

For the sedimentation, he implies that he is able to apply the rule of three according to the cumulated thickness of the Paleozoic, secondary and Tertiary strata – thus of all the fossiliferous grounds carrying the archives of the living 72.584 feet – and of the principle of actualism with regard to the speed of sedimentation of the Mississippi River of 600 feet in 100,000 years. It does not finalize the calculation. It seems likely that the value he would have set, 12,097,333 years or 12 million years, seemed insufficient to reinforce his theory.

According to him the erosion or "[denudation] probably offers the best evidence of the lapse of time" (ibid., p. 284). We can find the value he gives by always applying the two logics: the rule of three and the principle of actualism. His method of calculation is based on the specific case of a large Cretaceous rock dome in southern Great Britain, the Weald. This field study had already been made by Lyell in 1833 in his reference work, with a geological map to support it (Burchfield 1974, p. 302).

Darwin gives us an estimate of the erosion rate:

> Hence, under ordinary circumstances, I conclude that for a cliff 500 feet in height, a denudation of one inch per century for the whole length would be an ample allowance (Darwin 1859, pp. 286–287) and the measurements of the cliff under consideration. The distance from the northern to the southern Downs is about 22 miles, and the thickness of the several formations is on average about 1100 feet, as I am informed by Prof. Ramsay (ibid., p. 286).

6. Often, our collective memory tends to only remember the last edition, that is, the 1876 version, interpreted as being the most accomplished version of the researcher's reasoning.

However, he seems to conceive of an accelerated rate as a function of the height of the cliff "though no doubt the degradation of a lofty cliff would be more rapid from the breakage of the fallen fragments" (ibid.). This would lead the reader to consider an erosion rate of 2.2 inches per century, though Darwin does not give the exact calculation he made. The length of the cliff of the Weald, bitten little by little by the waves, being 22 miles[7], we should find the value only of 63,360,000 years, or, in other words, 63 million years. However, to find the precise value given by Darwin, we must assume in reverse that a higher cliff of 1,100 feet would take longer to erode at the rate of 1/2.2 inches per century.

Considering t = d:v then bringing back the same unit of measurement in inches:

t = 22 x (5,280 x 12): 1/2.2 = 3,066,624 centuries or 306,662,400 years

We can thus understand the attack launched in 1860[8] by the English geologist J. Phillips (1800–1870), then president of the Geological Society of London, who refuted this estimate in view of the erroneous mathematical reasoning made by Darwin with his demonstration of the Weald, as it was based on a geological postulate. He gave an estimate of 100 million years for the age of the Earth, which might have satisfied the naturalist anyway (Shaviv 2009).

Let us note that Darwin shows extreme caution with respect to his mathematical demonstration, which he knows can be criticized given the number of variables to be taken into account, to which he alerts us throughout the text – for sedimentation, the heterogeneity of the deposits of the same stratum according to the regions (nature and height), and the non-measurable sedimentary gaps, and for erosion he focused on a single seafront and differential according to the hardness of the rocks, the height of a cliff and the exposure to the tides. However, his aim is not to give an exact estimate of either the age of the Earth, nor of the time of stratigraphic deposits or even an erosion time, but to make us aware that we are dealing with relatively long periods of time, "in order to form some crude notion on the subject" and due to "the vain endeavor to grapple with the idea of eternity" (Darwin 1859, p. 285 and 286) which may have allowed the evolution of living beings through natural selection.

7. One mile is equivalent to 5,280 feet and one foot to 12 inches, which equals a total of 63,360 inches.

8. Phillips, J. (1860). *Life on the Earth: Its Origins and Succession*. Cambridge University Press. Quoted in Shaviv (2009, p. 23).

7.2.4. A problem, object of controversy between historical science and experimental science

The problem was taken up within a wider community in the 19th century by two opposing camps, that of the naturalists – mainly geologists – and that of the physicists with their two respective leaders, Darwin – supported by his privileged follower T. Huxley (1825–1895) – and W. Thomson (1824–1907) – better known as Lord Kelvin.

Thomson's reasoning is based on two presuppositions: an energy acquired only during the formation of the solar system and an entirely solid and rigid globe, following the example of Buffon. He considered different measurements of the geothermal gradient in the mines, on average an elevation of 1 degree per 30 meters of depth. He then applied the new thermodynamic laws of heat diffusion given by J. Fourier (1768–1830) to estimate a cooling time of the planet by conduction. He wrote that "we may suppose the consolidation to have taken place 98,000,000 years ago", a value within a proposed range of 20–400 million years (Thomson 1864, p. 160). Kelvin is bound by the unquestionable laws of heat conduction in a solid and can be satisfied with a short time.

Geological principles and biological mechanisms, however, required much more time, even an immeasurable amount of time, while the laws of physics allowed an estimation. On the one hand, among geologists, he rejected the "doctrine of uniformitarianism" (Thomson 1866) or the principle of actual causes defended by Lyell, an eminent geologist Darwin relied on, who assumed an infinitely long time. He estimates, in a uniformitarian framework, the following equivalence of quantity of heat lost by the Earth:

> [I]f this action had been going on with any approach of uniformity for 20,000 million years, the amount of heat lost out of the earth would have been about as much as would heat, by 100°Cent., a quantity of ordinary surface rock of 100 times the earth's bulk (ibid., p. 512).

This is inconceivable because the Sun and the Earth are finite bodies that can be estimated to be approximately 20 million years old. On the other hand, with respect to biologists, without contesting the idea of the evolution of living beings, he refused, for philosophical reasons, Darwinian selection based on the idea of chance (Smith and Wise 1989). This is well captured in a reconstructed historical dialogue where Thomas is given the floor and disagrees with Huxley:

> I am not opposed to evolution, but I am against natural selection because it leaves no room for the operation of design or a divine order in the evolution of life. For me design is just as much a principle of nature as the laws of thermodynamics. The idea of design and scientific laws go hand in hand (Stinner and Teichmann 2003, p. 224).

In addition to a different epistemic framework, it is interesting to note that there are two approaches to proof: one in historical science by the naturalist geologist mobilizing evolutionary, unique and contingent events; the other in experimental science by the physicist summoning laws to study a universal phenomenon of thermal conduction. Moreover, the authority of the physicist through their fame and influence[9], and the overvaluation of the proof through experimentation and mathematical demonstration could give rise to a momentary renunciation of the community of geologists. However, their doubt is present in the numerous areas relevant to the calculation of the estimated duration of the physicists: the initial temperature of the Earth, the thermal conductivity or average specific heat of the terrestrial materials, the temperature of fusion of the rocks of the terrestrial crust and finally the geothermal gradient (ibid.). Lyell, 20 years earlier, had already refuted the application of Fourier's mathematical model to the Earth:

> But his reasoning [to Fourier] seems to be confined to the conduction of heat through solid bodies; and the conditions of the problem are wholly altered when we reason about a fluid nucleus, as we must do, if it be assumed that the heat augments from the surface to the interior, according to the rate observed in mines. For when the heat of the lower portion of a fluid is increased, a circulation begins throughout the mass, by the ascent of hotter, and the descent of colder currents (1840, p. 437).

Subsequently, the two postulates on which his reasoning was based were invalidated by new discoveries at the end of the 19th century and at the beginning of the 20th century. The postulate of the only primordial energy source was refuted by the discovery of an internal heat source linked to radioactivity within magmatic rocks. In addition, the postulate of an entirely rigid globe, even though it is mostly solid, is refuted by discoveries in seismology, including the viscosity of the mantle that allows convection movements.

9. In 1866, Queen Victoria ennobled W. Thomson, who became Lord Kelvin and, in 1892, was made Baron Kelvin. He was a member and then President of the Royal Society from 1890 to 1895.

7.3. Epistemological insights and the problematization of the age of the Earth

Let us now see how the construction of the problem of the age of the Earth can be enlightened by epistemological reference points in order to show the didactic potential of the historical source that has just been studied in its contexts of production and reception in the 19th century.

7.3.1. *Short time versus long time: a major and recurrent obstacle in the history of science*

From an epistemological point of view, the acceptance of deep geological time has been a major and recurring obstacle in the history of science (Gould 1990; Boughamni 2019). How did the age of the Earth, believed according to dogma to only be a few thousand years old, become a few billion years thanks to scientific reasons?

In the 18th century, the empirical demonstration of Buffon (Comte de Buffon 1774) of the cooling time of iron balls of various sizes, was a first break from scripture:

> It would take about one hundred thousand years to cool down to the present temperature a globe of iron as big as the Earth, without counting the duration of the first state of liquefaction, which still pushes back the limits of time, which seems to flee and expand as we try to grasp it (Comte de Buffon 1775a, p. 36).

This new conception of a long time was attacked by numerous big players: Voltaire, frightened, preferred to remain in a permanentist framework of "remote times that frighten the imagination" (Voltaire 1768/1879, p. 153), and the Sorbonne wanted to maintain the legitimacy of a creationist framework (Comte de Buffon 1753). These modes of thought, of permanence of the current landscapes with their mountainous reliefs and their positions of the seas, and of an immaterial entity at the origin of living beings, were able to interfere to explain the history of the physical world in parallel with that of the living world (Hooykaas 1963).

The 19th century saw the confrontation of physicists who minimized the age of our planet under the authority of Kelvin (1895), carried by the laws of thermodynamics and naturalists who maximized it according to the *Principles of Geology*, founded by Lyell (1843–1848):

> But if the paleontologists have good reasons for demanding much greater times, I see nothing from the physicist's point of view which denies them four times the greatest of these estimates [1000, 400 or 500 million years] (Perry[10] 1895, p. 585).

It was only in the middle of the 20th century that Patterson, thanks to the discovery of radioactivity by Becquerel in 1896 and the periodicity of the radioactive elements by Rutherford and Soddy in 1902, validated the vertiginous age of the Earth sensed for a long time by geologists.

7.3.2. *From an assertoric knowledge to an apodictic knowledge*

The question of the age of the Earth has now been solved. The scientific community managed to reach consensual and factual closure: it is approximately 4.5 billion years old[11]. This was proposed in 1955 by C. Patterson and his collaborators (Patterson et al. 1955). It can be declared a simple assertion or proposition given as true. In this light, it risks being a false problem for the school community if its historical genesis is obscured and if the dogmatization of propositional knowledge prevails over the heuristics and logic of research (Fabre 2007). On the contrary, putting this empirical result into perspective with an immersion in the history of science will allow us to redefine the construction of the original problem: facts mobilized in many disciplines, divergent explanatory models between naturalists and physicists, and epistemic frameworks unique to each community.

Synthesis works in ELS didactics that stem from the theoretical framework of problematization (Orange 2012; Orange Ravachol 2012; Lhoste 2017) want to give a certain weight to the problem treated in class without rushing towards its resolution or even "the" solution. According to the theoretical framework of problematization (Fabre 1993; Fabre and Orange 1997) for scientific knowledge under construction, the nature of the arguments can be categorized between two registers in tension: the empirical register mobilizes facts summoned with regard to the problem studied, and the register of explanatory models explores conditions of

10. John Perry, for a time an assistant under Kelvin's authority, ended up doubting one of the postulates put forward by the latter, that of the rigidity of the Earth. The temperature in the depths of the Earth could be so high as to melt the rocks, which would allow a transfer of heat by convection, and consequently, an underestimation of the cooling time, which would imply a considerable underestimation of the age of the Earth.

11. The data of an age by absolute dating in geology, in spite of this qualification, always has a margin of uncertainty of a few million years, more or less.

possibilities as solutions to the problem worked on. Among this network of possibilities, a solution can emerge from a necessary reason. It then becomes apodictic knowledge; that is, knowledge demonstrated as being unable to be reasoned in any other way and that constrains the scientific exploration of explanatory models.

We assume that the teacher's interventions can be a determining factor in the students' appropriation of the problem. In fact, the teacher must give the students real space for oral and written expression in order to access their reasoning. Giving the students the full meaning of the scientific problem requires the teacher to accompany them according to the three dimensions of problematization: the position, definition and resolution of the problem (Fabre 1999) revisited by Crépin-Obert (2017):

– the enigma is formulated according to their own questioning;

– a debate is co-constructed by their sustained argumentation, putting in tension selected facts and explanatory ideas;

– the tests are understood as a test of a field of possibilities with new correlated facts.

The age of the Earth can become a reasoned solution for students if they question its validity in science by differentiating it from other values linked to beliefs, if they understand the various empirical methods of estimating a duration and if they become aware of past controversies about the issue. Furthermore, understanding the concept of deep time is necessary as a framework for all learning about events and phenomena in geology (Trend 1998; Orange Ravachol 2005).

The historical problem of the age of the Earth thus reveals the three dimensions of building a problem associated with the characteristics of a scientific activity, shown here according to its three decisive stages:

– The 18th century saw the structuring of the perimeters of science in relation to religious dogma, raising a new problem or enigma.

– The 19th century saw the development of an international debate between geologists and physicists, which can be identified as a true scientific controversy in the sense of Raynaud (2018, p. 40).

> [A] scientific controversy is characterized by the persistent and public division of several members of a scientific community, coalesced or not, who support contradictory arguments in the interpretation of a given phenomenon.

– The 20th century saw the resolution of the problem through the discovery of radioactivity, as well as through mathematical and empirical proof from samples of terrestrial rocks and meteorites.

7.3.3. *Learning objectives with an epistemological purpose refocused on the construction of a historical problem*

An in-depth study of the construction of this scientific knowledge surrounding the age of the Earth makes it possible to illustrate the criteria of scientific nature identified through the history of the ages of the Earth. These criteria can constitute learning objectives with an epistemological aim, in addition to the historical and scientific – notional and methodological – aims. The characteristics of scientific activity can thus be targeted as a teaching issue from the point of view of historical problematization. We have broken them down into five correlated specificities: (1) socio-historical knowledge anchored in a past cultural context (relations between science and society, interdependencies between science and beliefs); (2) evolving, rectified and therefore provisional knowledge (creativity, heuristic approach, technical progress, evolution of ideas through continuity or epistemological rupture); (3) common knowledge elaborated within a scientific community (objectivity by submission to the criticism of peers, debate); (4) explanatory and problematized knowledge built by logical inferences or argued conjectures (empirical registers and explanatory models, theories and scientific laws; facts – observations and experimentation – confronted in coherence with concepts); and (5) proven knowledge within the experimental and/or historical sciences (countering the myth of the unique scientific method).

From our point of view, the problem lies at the heart of those epistemological specificities that characterize reasoned scientific knowledge. We wish to overcome the tensions between notional or epistemological objectives, no doubt a remnant of the old dichotomy between an internalist history centered on the scientific domain, and an externalist history open to the social, political, economic, cultural and geographical context (Mathy 1997). Let us recall our orientation based on the Kuhnian paradigm, "universally recognized scientific discoveries which, for a time, provide a community of researchers with typical problems and solutions" (Kuhn 1962, p. 11) framed by "a set of values and techniques shared by a scientific group" (Lecourt 1999, p. 718).

In this study, we would like to focus on the construction of the geological deep time problem by giving meaning to some epistemological learning objectives inherent to the age of the Earth, according to the five specificities defined above:

1) As a scientific thought, this problem was initially marked by the difficulty to be opened and then maintained in the field of sciences, due to the interference with a religious register based on sacred texts, giving a dogmatic time in thousands of years.

2) From being perceived, it becomes constructed in rational knowledge problematized through the rectification (Bachelard 1938) and articulation of successive problems evolving through opening and closing, in accordance with the Popperian unfinished quest (Popper 1976) – the formation and cooling of the Earth, stratigraphic and then evolutionary paleontology, phenomena and durations of erosion and sedimentation, internal energy and terrestrial radioactivity.

3) Taken in charge by the deep argumentation of different key players, this explanatory and refutable common knowledge (Popper 1978) is stirred by a major controversy, characterized by its extension, its duration and its intensity between geologists – a long time in billions of years – and physicists – a short time in millions of years.

4) As an apodictic knowledge (Orange 2012), it has finally been identified within an evolutionary theoretical framework that puts in tension an empirical register and a register of explanatory models that make it necessary to think about long fossiliferous times.

5) If proof through mathematical measurement can be preferred as an established demonstration of geological deep time, the problem remains rooted in historical sciences by networking other forms of proof related to the relative chronology of singular events linked to the history of life and their contingency (Orange Ravachol 2005; Lecointre 2011; Gray 2014) – the superposition and thickness of fossiliferous strata, the diversity and specificity of fossil faunas by their geological stage, and the progressive analogy of fossil genera and species to current biodiversity.

In the following, we propose one of the methodological tools of problematization, a space of constraints, schematizing the argumentative dynamics by Darwin and Thomson, based on the analysis of the founding text and its contexts of production and reception, conducted in the first part. For each of the key players debating the problem of the age of the Earth in the 19th century, reason is constrained by the relationship between facts deemed relevant and explanatory models that have been used or newly constructed, either by the naturalist or the physicist. Each one develops the coherence of his reasoning within a federating theoretical or epistemic framework.

We emphasize that the argumentation created for each protagonist during this debate is based on very different registers – the empirical register and the register of explanatory models. We also evaluate how fundamental the theoretical framework, evolutionary or not, is in orienting the selection of data – geological and biological field observations, measurements, modeling – for a prediction of a long or short time. Darwin needed an immensely long time to support the model of natural selection that would generate the modifications of past life to the very great diversity of current fauna and flora, which he became aware of during his voyage around the world on the HMS Beagle.

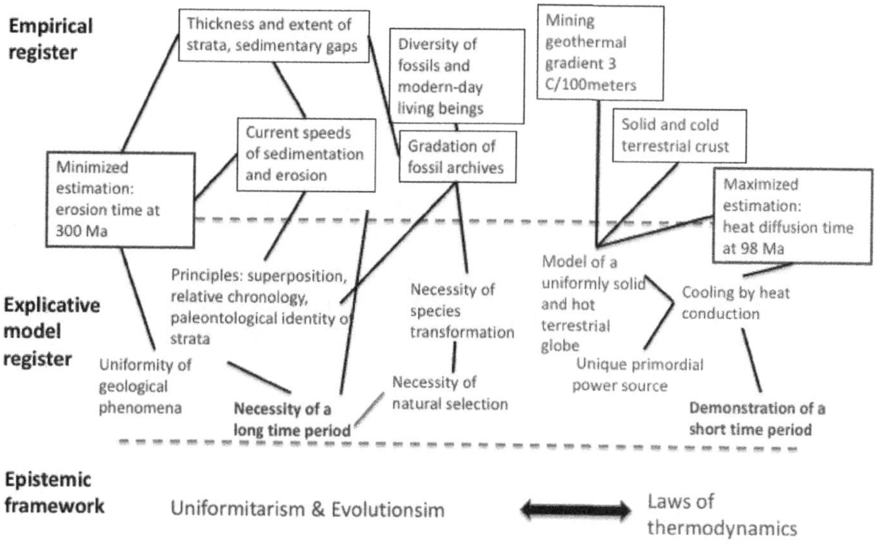

Figure 7.1. *Constraint space on the age of the Earth controversy between naturalist Darwin (1859) and physicist Thomson (1862). For a color version of this figure, see www.iste.co.uk/fortin/evolution.zip*

We have chosen to highlight the estimated durations by positioning them at the interface between the empirical register and the register of explanatory models. Indeed, they have a particular status in the reasoning of the scientific debate, being neither an established fact nor an assured explanatory model. These contrasted durations constitute the underlying problem in creating a plausible argument. In previous research, we have put forward the status of "conditional necessity", a logical consequence yet to be demonstrated in the causal explanatory system of stabilized necessities (Crépin-Obert 2010). We can bring this particular case closer to the status of "empirical necessity", introduced in modified schematizations of constraint space of "undiscovered empirical elements whose existence is made necessary, in a given theoretical framework" (Lhoste and Peterfalvi 2009, p. 85). They can be applied here to the historical problem of the depth of time of our planet. A good illustration of a constrained reason between the possible and the necessary (Canguilhem 1965), the immensity of geological time was a necessity conceived by the community of geologists in the 19th century, under the condition of subsequent validation by consensual proof, which was finally brought by physicists after they had been the detractors. With the discovery of radioactivity, the hypothesis that our planet is billions of years old became factual.

7.4. Problem and research questions

All the epistemological richness of constructing the historical problem of the age of the Earth allows us to envisage its high potential in science didactics. The possible success, as well as the difficulties of its appropriation in class, can be predicted by the affect linked to any question concerning origins – the formation of the universe and the Earth, the origin of life, the origin of species or the animal origin of the human species. Teachers are asked so many existential questions spontaneously by their pupils. Formulating rational and pedagogical answers is a societal challenge in the face of persistent opinions, such as that the Earth is flat, at the center of the universe and young (National Academy of Sciences and Institute of Medicine 2008). A historical approach to the advancement of scientific knowledge can help teachers address this challenge. However, to date, few studies have been focused on teaching practices that integrate elements of the history of science and their training (Maurines and Beaufils 2011; Crépin-Obert et al. 2021).

Within this study, our problem is the following: how and with what epistemological means do middle school science teachers take up the historical problem of the age of the Earth during a training course? This leads to several research questions:

– Is the age of the Earth still misunderstood and what opinions or beliefs are held today by students and their teachers?

– What understandings of the historical problem and what epistemological points of reference can be uncovered in teachers in training during problematization activities based on historical writings by Darwin?

7.5. Teacher training scenario and data collection

The guiding principle of the methodology of this research is the development, implementation and regulation over three consecutive years from 2017 to 2019 of an innovative continuing education course entitled “Interdisciplinarity and history of science in Cycle 4[12] of secondary school” in the academies of Créteil, Paris and Versailles. Offered within the framework of the IREM (Research Institute on Mathematics Teaching)[13] of Paris, the course was organized in alternating sessions – two days of acculturation to the history of science, followed by one day a few months later – and offered trainees the opportunity to initiate projects in their classes, either in physical sciences or in ELS. The training objective was to encourage the development of teaching practices that integrate historical elements by crossing different school disciplines – mathematics, physics–chemistry, ELS. Here, we will focus on the field of geology with the targeted problem of the age of the Earth over half a day of training – its shape, its atmosphere and the biosphere being seen in the other fields. We will specify the training device in three parts, of which the third constitutes the main collection of the research data.

Training time	Media used	Trainee activities	Data collected for research
1st	Six individual misconceptions of middle school students in their final year about the age of the Earth	– Initial collective misconceptions – Analysis in terms of prototypical alternative frameworks and obstacles – Personal position on this teaching method	Individual initial misconceptions of four classes of 9th grade students (83 students, aged 14–15)
2nd	Slide show in the form of a timeline about the history of the ages of the Earth	Knowledge of the scientific works that preceded and followed Darwin’s writings, situated in its historical, social and cultural context	

12. Children aged 13.

13. It should be noted that the IREM (*Institut de Recherche sur l’Enseignement des Mathématiques*) has been opening up for several years to research on the teaching of other disciplines, and particularly the sciences. New acronyms proposed in some academies (IRES, IREMS, IREM&S) reflect this evolution. Within the IREM group, formed in 2016, there is HistES (*Histoire et Enseignement des Sciences*, meaning History and Teaching of the Sciences), with the following team of trainers: R. Bosdeveix, S. Canac, P. Crépin-Obert, N. Décamp, C. Forgas, C. de Hosson, S. Javoy and F. Lalevée.

Training time	Media used	Trainee activities	Data collected for research
3rd	Excerpt from Darwin's "On the vast lapse of time, as inferred from the rate of deposition and denudation" (1859)	– Listening to what is read aloud, reactions and questions – Four activities to choose from in small groups – Presentations of the written records	Trainees' writings and speeches on the five activities

Table 7.1. *Description of the overall training system*

7.5.1. *First training period: emergence and collection of initial representations on the age of the Earth*

As the age of the Earth is a completely new topic compared to the different school programs that it followed, the spontaneous answers of the trainees are explored in the introduction of the course and in a collective mode. They are questioned on their answers a priori concerning the age of the Earth, the method of proof and the century of resolution.

Regarding this first approach, some contrasting misconceptions of middle school students are presented to them for succinct analysis in terms of major alternative frameworks and epistemological obstacles. Collected during a previous pedagogical experiment[14], with the goal of using them as a support during classroom debate, the misconceptions were derived from an individual questionnaire carried out in four 9th grade classes in 2016. The first question concerning the age of the Earth was: "*According to you, how old is the Earth?*" (Q1). Students were then asked about the methods used to gain this knowledge through three open-ended questions: "*How do you know?*" (Q2), "*Suggest clues, methods, or techniques to demonstrate this*" (Q3) and "*Are all rocks on Earth the same age? Justify this by giving scientific evidence*" (Q4). Finally, a timeline (Q5) that covered the formation of the Earth to the present day had to be completed in a limited time by proposing five biological and/or geological events.

14. Special thanks to ELS teacher C. Forgas, for participating in the development of the pedagogical sequence with the researcher, and its realization in 2015–2017 at the Pierre Brossolette middle school, REP (Priority Education Network) of Villeneuve-Saint-Georges in the Créteil academy.

7.5.2. *Second training period: plurality of methods on the ages of the Earth during three centuries*

In the form of a lecture, a synthesis explained the different actors who have taken up the problem of the age of the Earth, giving several ranges depending on the variety of methods they employed (De Wever 2005; Gindras 2009; Krivine 2011b, 2017; Candel and Jouanna 2017; Lemarchand 2017–2018):

– The biblical age of J. Ussher in 1650 was taken up by N. Lenglet Du Fresnoy in 1729, based on the genealogical demonstration of the ages of the patriarchs since the Creation of the world from the exegesis of the book of Genesis – an estimation of 4,004 years or 4,700 years was given.

– The cooling age: the first scientific questioning on the age of the Earth, supported by an experimental method, is carried out by the naturalist Buffon (Comte de Buffon 1774, 1775a, 1775b). He deduced this by extrapolating a cooling time for the Earth then for the solar system and the planets – the ages of 96,670 years then 74,832 years precisely were published. This method was taken up again by the physicist Thompson between 1862 and 1897 by relying on the universal laws of Fourier – 20–400 million then 20–40 million years were proposed.

– The age of ocean salinity and river flow by J. Joly in 1899 – 90 million years.

– The relative age of fossils and the time of sedimentation and erosion: B. de Maillet proposed 2 billion years in 1748. This method was also pursued by Buffon in a manuscript – he estimated it to be a few million to 3 billion years, then taken up again by Darwin in 1859. We have chosen to study this last text in particular, and we will explain this choice in the following section.

– Finally, following the discovery of radioactivity, the absolute age of the atom was studied. This is a form of dating that can be used to date the oldest terrestrial rocks, i.e. meteorites. Patterson (1956) thus demonstrates using three independent radiometric methods the conjecture of a telluric age of 4.55 billion years, plus or minus 70 million years for experimental margin of error.

7.5.3. *Third training period: study of a historical source and collection of teachers' productions*

This final stage is structured around a text from the first edition of Darwin's major work *On the Origin of Species* (2013 (first published in 1859)[15]), and the

15 The trainees worked on the version that was translated into French, presented and annotated recently by T. Hoquet.

historical and epistemological issues that have been presented above. From a didactic point of view, the exploration of other major historical sources that could potentially be used in class (Buffon, Kelvin, Rutherford, Patterson) shows that the transposition of Darwin's text is much easier for a middle school 9th grade class. A recent French translation, a syntax accessible to middle school students, even though sometimes the rhetoric or the metaphors require further explanation, a clarity of demonstration supported by notions acquired in the 7th grade (aged 12–13) on sedimentary phenomena, and the hazardous hypotheses of the rule of three constitute an interesting and affordable attempt in cycle 4 of middle school, including that of the scientifically founded sedimentation time.

"Texts and context" could be the adage of a historian's method, emphasizing the importance of studying a primary source, cross-referencing it with other sources of their time and putting it into context to best understand the thinking of a scientist. We thus wanted to immerse the trainees in authentic activities of analyzing a chosen primary source. After reading aloud and commenting on the chosen extract, we tested its reception as a didactic device with five activities[16] carried out by the trainees: reactions and questions to the first reading in a collective phase (A1); then, in groups, following different studies, constructing a diagram of the scientific community's interactions with Darwin (A2); organizing a diagram of the researcher's demonstration based on the complementary sedimentary phenomena of sedimentation and erosion (A3); for these phenomena, testing of the reproducibility of the author's calculations, which gives an estimate of 300 million years for the erosion of the Weald (A4); and carrying out a comparison of the extract in two editions from 1859 and 1873 (A5).

The trainees' group work was collected and the various moments of collective exchanges were recorded, including their reactions after the first reading and their presentations of the five activities, supported by the discussions triggered within the group of trainees.

7.6. Results

7.6.1. *The same heterogeneity in the spontaneous responses of the trainees and the misconceptions of the pupils in 9th grade*

Within the groups of trainees – mostly teachers of physical sciences and teachers of ELS, as well as some teachers of mathematics – from 2017 to 2019, the exact

16. The activities are numbered in relation to the materials and the trainees' speeches; they will be recalled in different parts of the analysis of the results and in Table 7.1.

answer of the age of the Earth is far from unanimous – ages ranging from 5,000 years to 6 billion years are reported by the trainees. Eclectic methods are given: in reference to different scientific fields (paleontology, geology, stratigraphy) – or even religious fields (the Bible); by the objects studied (rocks, strata, fossils, comets, asteroids and the Sun) or by the given reasoning (cooling, radioactivity including the recurrent C_{14} or Rb/Sr, simultaneous formation of the Earth and Sun, the size of layers and sedimentation speed). Finally, the various theories of the 18th, 19th and 20th centuries are considered.

The researcher's preliminary analysis of the initial framework for students in 9th grade shows an even greater diversity. The figure below shows two students' answers, theirs being representative of the variety of responses obtained and presented to the trainees.

Student A

1. **How old do you think the Earth is?**
   I think that the Earth is more than 75 million years old, a bit older than the Jurassic era.
2. **How do you know this?**
   I know because dinosaurs lived 75 million years ago (I think) and that the Earth was formed before that
3. **Provide indications, methods and techniques to show this**
   We can show this by adding the duration of each period of history such as pre-history, antiquity, the Middle Ages, etc.
4. **Are all of the Earth's rocks the same age? Justify your answer by providing scientific evidence**
   No rocks aren't the same age because since the creation of the Earth, rocks renew themselves by way of volcanic eruptions, which frees lava and by heating solidifies and creates new rocks

Student B

1. **How old do you think the Earth is?**
   In my opinion, the Earth is 10 billion years old
2. **How do you know this?**
   The first civilisations date back to the 18th millennial year BC
3. **Provide indications, methods and techniques to show this**
   To demonstrate the age of the Earth, we can study deeper rocks or look in religious books (the Bible, Koran, Tora).
   To choose between science and belief.
4. **Are all of the Earth's rocks the same age? Justify your answer by providing scientific evidence**
   All the rocks aren't the same age, the more rocks are found at a certain depth in the Earth, the older they are.

Figure 7.2. *Ninth grade middle school students' work on the age of the Earth*

Among the 83 students surveyed, the heterogeneity of their answers ensured a fruitful debate in class. Very different factual ideas are given in the first question:

"I think we can't know"; "the Earth is not old"; "the Earth is 2,016 years old"; or various lengths of elapsed time that range from centuries, thousands, millions (Student A) or billions of years (Student B). Divergent explanatory ideas affiliated with science are given through questions 2, 3 and 4: observation of superimposed layers, relative chronology index (students A and B), the use of fossils, including extinct species such as dinosaurs (student A) or prehistoric humans, analysis of radioactive elements in rocks centered on $Carbon_{14}$, drilling the Earth to its core, documentation or interviews with a scientist. However, from the outset, the field of possibilities is widened outside the scientific register by numerous answers: the current calendar as a reference point for everyday life, prehistoric and historical reference points (pupil A), the over-valuation of technology and the media, and reference to religious texts (pupil B) or to one's own opinion.

In both situations, in training and in class, the need to identify this knowledge is confirmed. Going back to the sources of its construction, posed in the 18th century and validated recently in the middle of the 20th century seems, according to us, to be of the most demonstrative methods.

7.6.2. *Personal conceptions in the epistemology of science out of step with the activity of a scientist*

An analysis grid is created *after* the trainees have produced their work. Modes of thought, at odds with the reality of the construction of knowledge on the age of the Earth, are identified. Mirroring the emerging obstacles of a naive epistemology (column 1 of Table 7.2), characteristics of an expert epistemology (column 2 of Table 7.2) are targeted as training objectives. These different benchmarks could be marked out according to the two axes of study developed in the previous sections. On the one hand, the practice of the history of science and historical culture allows for a reasoned immersion and distancing from an archive. On the other hand, the theoretical framework of the problematization is rooted in the philosophy of science and in the epistemology of science, shedding light on the criteria of scientificity in the construction of scientific problems and providing a critical look at the construction of knowledge: mostly, the French filiation of a historical epistemology is carried by Duhem, Bachelard, Canguilhem and Koyré (Jullien 2020); in addition, a few English-speaking actors, such as Popper and Kuhn, are also used as references.

The trainees' reception and analyses of Darwin's text, guided by the given tasks, reveal their personal conceptions of a problem worked on in the historical sciences in relation to the experimental sciences: the unique scientific method, the increased importance of observed facts and the expected notions of proof and truth. Moreover, the trainees take as a reference point their own teaching approach to understand this historical problem. The reception of the text through the activities carried out (reactions, written productions and presentations) allowed us to identify certain significant elements of an epistemology that we describe as naive when it comes to science (column 1 of Table 7.2), insofar as it is far from the historical problematization carried out by scientists.

The exchanges that took place after the presentations enabled them to distance themselves from the subject matter and move towards a more expert epistemology (column 2 of Table 7.2): a plural and heuristic approach, a coherent argumentative system between constructed facts and explanations and a theoretical explanatory framework supported by a set of facts, principles or laws and tested hypotheses – past, present and also future ones. Science is not a finite knowledge and its path is marked by errors, invalidated hypotheses and uncertainties.

1. Trainees' writings (W) and discourse (D), significant for a naive epistemology, later on shared	2. Discussions with trainees towards an expert epistemology
Method format or plural approach	
Myth of the single scientific method according to a constant and linear succession of steps: "*PHO – problems, hypotheses, observations*" (W-A3) and over-valuation of experimental evidence: "*He doesn't do measurements; he picks up what interests him from his reading*" (D-A1)	**A plurality of evidence-based approaches** in experimental and historical sciences
Over-valued facts	
A solely positivist or inductivist vision of science: "*He is in confirmation bias so he chooses data that support his theory*" (D-A1) "*There is a sense that he is integrating the findings that interest him only*" (D-A5)	**The facts are constructed**, not given. Selected or provoked by the researcher, they are confronted with the explanatory model according to different approaches
Confusion about the nature of the arguments: "*The term 'illustration' is used to describe a model by James Croll on which Darwin relies*" (W-A2) "*He says that in our sedimentary layers there are gaps [...]. This is a fact that he does not explain [...]. Today we know that sedimentary gaps do exist*" (D-A3)	Clearly differentiate **the argumentative nature of facts and explanations**
A fact as proof in itself: "*On the fact that erosion takes time we distinguished two things: what he sees, what he really sees, what acts as evidence and what is only an explanation, what he imagines, what he thinks without really having evidence*" (D-A3)	**Scientific proof** is built by putting in tension arguments from an empirical register and arguments belonging to the register of explanatory models

Personal belief or scientific framework of thought	
Conviction in the sense of a simple personal opinion: *"It reminds us a little of the idea of convincing and we find the idea that if we write well everyone will agree"* (D-A1)	A **theoretical framework** (in this case evolutionary) directs all the researcher's thinking; later on, this framework can be called a paradigm
Negative relationship with uncertainties, against an expected incontestable "truth": *"I also agree with this idea (conviction), because when we read, each time he makes a lot of approximations: he says 'oh, we will consider the size of the layers, we will consider the speed, we will rely on the strata which are like this and like that' ... and all of a sudden he arrives at this rate, let's say 300 million years"* (D-A1)	**Uncertainties** are an integral part of the explanatory models developed, alerting to the degree of control of the systemic set of parameters
Discrediting or even rejecting Darwin's demonstration by considering only the mathematical error: *"T1: We didn't quite understand why especially he made that assumption there. Hence the question mark"* *T2: "We would even be of the opposite opinion because if we erode the lower part and that collapses it makes scree that will be protective, it makes a breakwater at the foot of the cliff, therefore that will go less quickly. If we are not allowed to collect pebbles on the beaches of Normandy, it is to protect the cliffs [...]"* *T2: "In fact, we don't understand what he wants"* *T1: "In conclusion, we did not understand anything"* (D-A4)	Understand a **heuristic approach** that explores a field of possible hypotheses, some of which will be invalidated
Teaching, communication or research in progress	
Taught science versus laboratory science: *"I found that it emphasized the fact that he was trying to convince, so he obviously explains but the intention is to convince, whereas our approach when we have a course is to explain, I do not intend to convince"* (D-A1)	An **explanatory system for ongoing research** that is distant from a communication or teaching model
Representation of the scientist's activity held up to current scientific activity *"I find that it brings a little bit more science to it by making diagrams, or doing calculations. It brings us to the idea that we are closer to something real"* (D-A1)	**The standards for publishing scientific research have evolved** over time

Table 7.2. *Trainees' conceptions and contrasting epistemological reference points*[17]

17. We recall here the coding of the five activities carried out by the trainees: (A1) reactions to and questions on the first reading; (A2) construction of a diagram showing the interactions of the scientific community; (A3) schematization of the scientist's reasoning based on sedimentary phenomena; (A4) testing of the reproducibility of the author's calculations for

The trainees recurrently described Darwin's approach to the search for geological time as a "conviction" (column 1 of Table 7.2): the selection of field data, thinking framed by the evolution of living organisms, the manipulation of numerical values with numerous estimates and the absence of diagrams. They discredit this approach, which does not conform to a scientific approach, as they see it, focused on experiments and measurements.

Moreover, several critical points relating to the construction of the problem were identified by the trainees (column 2 of Table 7.2) depending on the activities carried out: the importance and the complexity of the interactions within a multidisciplinary scientific community, in spite of the uncertainties, the extreme precision of a monstration by coherence of the arguments without recourse to the experiment, the weight of a theoretical framework which guides reasoning, the non-reproducibility of the advanced mathematical proof and finally, the reliability of the various editions of a publication.

Ultimately, Darwin's account is quite characteristic of the epistemological foundations of an approach rooted in historical science. By focusing on the scientific problem, certain learning objectives in the epistemology of science make sense: controversial hypotheses within a scientific perimeter within a community of geologists and physicists, rebounding from successive problems giving increasing scope to geological time, mathematical evidence privileged in conjunction with other evidence in historical science related to relative chronology and the importance of the evolutionary framework.

7.7. Conclusion and outlook

We retain three salient points concerning the results of this study through the teachers' reception of a didactic device based on the history of science, bringing out the interests they project and the limits to their comprehension.

First, we note a closeness and great diversity of representations among middle school teachers and their students about the current scientific assertion of the age of the Earth. The numerous spontaneous ideas outside the scientific field show the permanent struggle against the common opinion. If the scientific problem is to be devolved to the students by conducting a fruitful classroom debate based on the prototypical alternative frameworks, this can only be achieved if the perimeter

estimating the durations of sedimentary phenomena; and (A5) comparison of the excerpt in two editions of *The Origin of Species* (1859 and 1873).

between science and beliefs is clarified beforehand. This will be the subject of a later study.

As a second point, we value the interest of an in-depth analysis of a founding text by Darwin to truly access the reasons debated of a historical problem in the past. Darwin's demonstration is to be appreciated from a heuristic point of view in this period of the mid-19th century when awareness surrounding deep time, initially defended by geologists, will become effective among the scientific community. This exploration allows us to get to the heart of the problem: the inference of deep time and the biological mechanism of natural selection. Moreover, this historical and epistemological study of a primary source is meant to complement the syntheses tracing the history of ideas on the age of our planet, through a narrative in the form of a frieze supported by dates, authors and discoveries that is intended for teachers (Deparis 2001; Krivine et al. 2007) or the general public (Richet 1999; Krivine 2011a; Ameisen 2016). Note that Darwin is explicitly mentioned in the list of scientists prescribed for science education in the 11th grade (aged 16–17) in the new high school programs[18].

Finally, teaching sessions that integrate historical elements open up learning perspectives based on the epistemology of science. However, to perceive this ambition in one's teaching presupposes an awareness of the construction of a problem in historical science or in experimental science. However, we are sometimes challenged in teacher training by the gap between the epistemological reference points of the scientist's activity and the teachers' representations of science. Identifying and revaluing historical sciences, measuring their specificities and their articulation with experimental sciences, are challenges to be taken up in official prescriptions and in the training of ELS teachers (Crépin-Obert et al. 2020).

7.8. References

7.8.1. *Contemporary references*

Ameisen, J.-C. (2016). Darwin et le problème de l'âge de la Terre [Online]. Available at: https://www.franceinter.fr/emissions/sur-les-epaules-de-darwin.

Bachelard, G. (1938). *La formation de l'esprit scientifique*. Vrin, Paris.

18. "*Histoire, enjeux et débats*" ("History, issues and debates") from Bulletin official No. 1 of January 22, 2019 (*Enseignement scientifique du tronc commun – Classe de première*): "Some major steps in the study of the age of the Earth: Buffon, Darwin, Kelvin, Rutherford. Modalities of building a scientific approach to a controversial issue to arrive at a stabilized result."

Boughanmi, Y. (2019). Quelques obstacles à l'essor du temps profond en sciences de la nature. *Bulletin d'histoire et d'épistémologie des sciences de la vie*, 26, 97–122.

Burchfield, J.-D. (1974). Darwin and the dilemma of geological time. *Isis*, 65(3), 300–321.

Candel, S. and Jouanna, J. (eds) (2017). Histoire de l'âge de la Terre. *5 à 7 de l'Académie des sciences. Histoire et philosophie des sciences*, 17 January [Online]. Available at: https://www.academie-sciences.fr/fr/Seances-publiques/age-terre.html.

Canguilhem, G. (1965). *La connaissance de la vie*. Vrin, Paris.

Crépin-Obert, P. (2010). Construction de problèmes et obstacles épistémologiques à propos du concept de fossile : étude épistémologique comparative entre des situations de débat à l'école primaire et au collège et des controverses historiques du xviie au xixe siècle. PhD Thesis, Université de Nantes.

Crépin-Obert, P. (2017). Pratique de débat et problématisation en paléontologie. Le problème de la filiation entre ammonite et escargot en CM2. In *La pratique de l'enseignant en sciences : comment l'analyser et la modéliser ?*, Bächtold, M., Boilevin, J.-M., Calmettes, B. (eds). Presses Universitaires de Louvain, Louvain-la-Neuve.

Crépin-Obert, P., Dessart, F., Paulin, F., Orange Ravachol, D. (2020). Regards croisés sur trois problèmes historiques enseignés en Sciences de la vie et de la Terre. Symposium "Interactions entre enseignement des SVT et histoire : Histoire de la vie et de la Terre, sciences historiques, histoire des sciences". In *Actes du Congrès de l'AREF*, Anna, B. and Lucie, L. (eds). Bordeaux, 3–5 July 2019, 352–363.

Crépin-Obert, P., Canac, S., Chorlay, R., Roux-Goupille, C., Décamp, N., Javoy, S., Kermen, I. (2021). Analyse de séances "ordinaires" intégrant l'histoire des sciences. Regards croisés sur quatre pratiques enseignantes en sciences et en mathématiques. In *Actes des 11e rencontres de l'ARDiST*, Marie-Noëlle, H. and Corentin, P. (eds). Brussels, 31 March–2 April 2021, 55–62.

De Wever, P. (2005). Temps de l'Homme, temps de la Terre. In *La mesure du temps dans l'histoire de la Terre*, Labrousse, L., Raymond, D., Schaaf, A., De Wever, P. (eds). SGF – Vuibert, Paris.

Deparis, V. (2001). Quel âge a la Terre ? [Online]. Available at: http://planet-terre.ens-lyon.fr/article/histoire-age-Terre.xml.

Fabre, M. (1993). De la résolution de problème à la problématisation. *Les sciences de l'éducation – Pour l'Ere nouvelle*, 4–5, 71–101.

Fabre, M. (1999). *Situations-problèmes et savoir scolaire.* Presses universitaires de France, Paris.

Fabre, M. (2007). Des savoirs scolaires sans problèmes et sans enjeux. La faute à qui ? *Revue française de pédagogie*, 69–78.

Fabre, M. and Orange, C. (1997). Construction de problème set franchissement d'obstacles. *Aster*, 24, 37–57.

Gindras, Y. (2009). Mais quel âge a donc la Terre ? *La Recherche*, 434, 92–94.

Gould, S.J. (1990). *Aux racines du temps*. Grasset, Paris.

Gray, R. (2014). The distinction between experimental and historical sciences as a framework for improving classroom inquiry. *Science Education*, 98(2), 327–341.

Hooykaas, R. (1963). *Natural Law and Divine Miracle: The Principle of Uniformity in Geology, Biology and Theology*. E.J. Brill, Leiden.

Jullien, V. (2020). *Ce que peuvent les sciences : une enquête*. Éditions matériologiques, Paris.

Krivine, H. (ed.) (2011a). Première partie. Son âge. In *La Terre, des mythes au savoir*. Cassini, Paris.

Krivine, H. (2011b). Histoire de l'âge de la Terre. *Image de la physique*, CNRS, 44, 15–20.

Krivine, H. (2017). Comment la Terre a-t-elle pu vieillir de plus de quatre milliards d'années en quatre siècles ? *C.R. Géoscience*, 349, 91–95.

Krivine, H., Campi, X., Périllat-Mercerot, M.-L. (2007). L'âge de la Terre. SCÉRÉN-CNDP [Online]. Available at: http://cndp.fr/themadoc/histoiredelaterre/presentation.htm.

Kuhn, T.S. (1962). *The Structure of Scientific Revolutions*. The University of Chicago Press, IL.

Lecointre. G. (2011). Récit de l'histoire de la vie ou de l'utilisation du récit. In *Les mondes darwiniens. L'évolution de l'évolution*, Heams, T., Huneman, P., Lecointre, G., Silberstein, M. (eds). Éditions Matériologiques.

Lecourt, D. (ed.) (1999). *Dictionnaire d'histoire et philosophie des sciences*. Presses universitaires de France, Paris.

Lemarchand, F. (2017–2018). La Terre, une planète qui ne fait pas son âge. *La Recherche, Hors-série n°24. Les plus grandes controverses scientifiques*, 30–34.

Lhoste, Y. (2017). *Épistémologie et didactique des SVT : langage, apprentissage, enseignement des sciences de la vie et de la Terre*. Presses universitaires de Bordeaux, Pessac.

Lhoste, Y. and Peterfalvi, B. (2009). Problématisation et perspective curriculaire en SVT : l'exemple du concept de nutrition. *Aster*, 49, 79–108.

Mathy, P. (1997). *Donner du sens au cours de sciences*. De Boeck, Paris.

Maurines, L. and Beaufils, D. (2011). Un enjeu de l'histoire des sciences dans l'enseignement : l'image de la nature des sciences et de l'activité scientifique. *RDST – Recherches en didactique des sciences et des technologies*, 3, 271–305.

National Academy of Sciences and Institute of Medicine (2008). *Science, Evolution, and Creationism*. The National Academies Press, Washington.

Orange, C. (2012). *Enseigner les sciences. Problèmes, débats et savoirs scientifiques en classe*. De Boeck Education, Brussels.

Orange Ravachol, D. (2005). Problématisation fonctionnaliste et problématisation historique en sciences de la Terre chez les chercheurs et chez les lycées. *Aster*, 40, 177–204.

Orange Ravachol, D. (2012). *Didactique des sciences de la vie et de la Terre : entre phénomènes et évènements*. Presses universitaires de Rennes, Rennes.

Popper, K.R. (1976). *Unended Quest: An Intellectual Autobiography*. Collins, London.

Popper, K.R. (1978). *Conjectures and Refutations: The Growth of Scientific Knowledge*. Routledge and Kegan Paul, Henley.

Raynaud, D. (2018). *Sociologie des controverses scientifiques*. Éditions matériologiques, Paris.

Richet, P. (1999). *L'Age du monde. A la découverte de l'immensité du temps*. Le Seuil, Paris.

Shaviv, G. (2009). The controversy about the age of the Earth. In *The Life of Stars: The Controversial Inception and Emergence of the Theory of Stellar Structure*, Shaviv, G. (ed.). Magnes, Heidelberg.

Smith, C. and Wise, N. (1989). *Energy and Empire. A Biography Study of Lord Kelvin*. Cambridge University Press, Cambridge.

Stinner, A and Teichmann, J. (2003). Lord Kelvin and the age-of-the-Earth debate: A dramatization. *Science & Education*, 12, 213–228.

Trend, R. (1998). An investigation into understanding of geological time among 10 and 11 year old children. *International Journal of Science Education*, 20(8), 973–988.

7.8.2. *Historical sources*

Buckland, W. (1824). Notice on the Megalosaurus or great Fossil Lizard of Stonesfield. *Transactions of the Geological Society, Second Series*, 1, 390–396.

Comte de Buffon, G.-L.L. (1751/1753). Lettre de MM. Les Députés et Syndic de la Faculté de Théologie, à M. de Buffon ; Réponse de M. de Buffon ; Seconde Lettre. *Histoire Naturelle, générale et particulière*, Volume 4. Imprimerie royale, Paris, V–XVI.

Comte de Buffon, G.-L.L. (1774). Premier mémoire. Expériences sur le progrès de la chaleur dans les corps. *Histoire naturelle, générale et particulière, servant de suite à la Théorie de la Terre et d'introduction à l'histoire des minéraux. Supplément*, Volume 1, Partie expérimentale. Imprimerie royale, Paris, 143–172.

Comte de Buffon, G.-L.L. (1775a). Huitième mémoire. Expériences sur la pesanteur du feu et sur la durée de l'incandescence. *Histoire naturelle, générale et particulière, servant de suite à la Théorie de la Terre. Parties expérimentale et hypothétique. Supplément*, Volume 2. Imprimerie royale, Paris, 1–36.

Comte de Buffon, G.-L.L. (1775b). Partie hypothétique. Premier mémoire. Recherches sur le refroidissement de la Terre et des planètes. *Histoire naturelle, générale et particulière, servant de suite à la Théorie de la Terre. Parties expérimentale et hypothétique. Supplément*, Volume 2. Imprimerie royale, Paris, 361–515.

Darwin, C. (1859). *On the Origin of Species, by Means of Natural Selection or the Preservation of Favoured Races in the Struggle for Life*, 1st English edition. John Murray, London.

Darwin, C. (ed.) (1869). On the imperfection of geological record. In *The Origin of Species by Means of Natural Selection, or the Preservation of Favoured Races in the Struggle for Life*, 5th edition. John Murrray, London.

Darwin, C. (ed.) (1872). On the imperfection of geological record. In *The Origin of Species by Means of Natural Selection, or the Preservation of Favoured Races in the Struggle for Life*, 6th edition. John Murrray, London.

Darwin, C. (ed.) (1873) IX : Imperfection des archives géologiques. *L'origine des espèces au moyen de la sélection naturelle ou de la lutte pour l'existence dans la nature*, 5th and 6th editions translated and annotated by J.-J. Mouliné, Reinwald, C. (ed.). Cie, Paris [Online]. Available at: http://darwin-online.org.uk.

Darwin, C. (ed.) (2013). IX : Sur l'imperfection des archives géologiques. In *Sur l'origine des espèces au moyen de la sélection naturelle ou la préservation des races favorisées dans la lutte pour la vie*, translated and annotated into French by T. Hoquet. Éditions du Seuil, Paris.

Kelvin, L. (1895). The age of the Earth. *Nature*, 1323(51), 438–440.

Lyell, C. (ed.) (1840). XIX: Causes of earthquakes and volcanos. In *Principles of Geology: Or, The Modern Changes of the Earth and its Inhabitants Considered as Illustrative of Geology*, 6th English edition, Volume 2. John Murrray, London.

Patterson, C. (1956) Age of meteorites and the Earth. *Geochimica et Cosmochimica Acta*, 10(4), 230–237.

Patterson, C., Tilton, G., Inghram, M. (1955). Age of the Earth. *Science*, 121(3134), 69–75.

Perry, J. (1895). The age of the Earth. *Nature*, 1329(51), 582–585.

Thomson, W. (1864). On the secular cooling of the earth. *Transaction of the Royal Society of Edinburgh*, 23, 157–169 [Accessed 28 April 1862].

Thomson, W. (1866). The "Doctrine of uniformity" in geology briefly refuted. *Proceedings of the Royal Society of Edinburgh*, 5, 512–513 [Accessed 18 December 1865].

Voltaire (ed.) (1768/1879). Des singularités de la nature par un académicien de Londres, de Bologne, de Pétersbourg, de Berlin, etc. *Mélanges VI, Œuvres complètes de Voltaire*, New edition. Garnier frères, Paris.

8

The "Conceptual Landscape" of Evolution: A Possible Instrument for Training of Earth and Life Sciences Teachers

Fabienne PAULIN
S2HEP, Université de Lyon 1, France

8.1. Introduction

The theory of evolution is the unifying framework within which knowledge about the biosphere is constructed and developed. Its teaching is a main objective in terms of learning about the fundamental concepts that structure the scientific fields of biology and, in part, geology. However, this is made challenging because of a recurrent difficulty in overcoming the well-documented obstacles among students of ontological finalism, Lamarckian transformism, predominant and ad hoc natural selection and the invisibility of contingency (Fortin 2000, 2009, 2011; Aroua 2002; Olander 2008; Zabel and Gropengießer 2011; Gobert 2014)[1]. This is the paradox of evolution (and a didactic challenge) when it comes to teaching it; an apparent "narrative" simplicity with "beautiful stories" of adaptation hiding, however, the

1. Some authors have conducted research on the representations of biology teachers on evolution but focused on the impact of their religious beliefs on teaching (Munoz et al. 2007), which is not the subject here.

The Evolution of Life: Teaching, Learning and Training,
coordinated by Corinne FORTIN and Julie GOBERT.

counterintuitive face (Kampourakis 2020) of the Darwinian explanation: an "adaptation without intention".

The guiding principle of the work that we have been carrying out for several years is that of establishing the conditions of possibility for effective teaching of the sciences of evolution. By effective, we mean teaching that leads to a problematized knowledge, i.e. mobilizing a dynamic network of scientific concepts, articulated in a historical perspective, allowing for in-depth work on evolutionary problems (Fabre and Orange 1997; Orange 2012; Orange Ravachol 2012). The construction of such knowledge is a wide-reaching aim and the work presented here does not claim to respond to this vast project, but it instead explores certain epistemological aspects of the knowledge mobilized by certain actors: teachers of Earth and Life Sciences (ELS), evolutionary researchers and students in the high school senior science (HS) class aged 16–18 who have been taught evolution. More broadly, this work takes place in a research framework that combines didactic and epistemological approaches and whose issues are explained and developed in depth in Yann Lhoste's book *Epistemologie et didactique des SVT* (Lhoste 2017).

The objective is to reveal the specific frameworks of thought of each of the groups studied, in which the various protagonists work on evolutionary problems. We thus seek to find out which concepts are the most present and/or the most salient ones mobilized by each of the groups. These frameworks of thought are used here as references shared within a community that, we believe, ultimately determine the content of the learning activities, as well as the conditions of their appropriation. They are also a plausible gateway to the epistemology implemented in the teaching of evolution that we are seeking to characterize.

To bring out these common references on evolution, we will mobilize a tool from sociology: the association network, which allows access to the "social or professional representations"[2] of a group constituted on a given theme, through a statistical treatment of the items[3] collected (Abric 1994, 2003; De Rosa 2003). This tool has already been used in the field of "education for" and "sustainable development" (Barthes and Alpe 2016). We use it, here, in a modified form, adapted to our research questions on the teaching of evolution. It allows us to bring out the representations of a group of people that, according to Abric, are structuring for this group. We have modified some of the modalities in order to collect additional information and construct more explicit links between the concepts of evolution

2. The concept of "social representation" was first proposed and explained in 1961 by Serge Moscovici in the field of social psychology.

3. The items are the words or expressions that the participants will use in the survey. We will keep this name in the analysis.

mobilized by the participants in the survey. What we call "conceptual landscapes"[4] are a cartography of the semantic networks relating to the evolutionary concepts collected from the different groups studied: researchers, teachers and students.

We also work within the framework of an updated theory of evolution that has been progressively enriched, since the 1970s, by new knowledge and that mobilizes a more diversified epistemology of evolutionary scenarios. The different analyses produced allow us to integrate these updates in the world of French secondary school teaching.

The results are rich in information and allow us to refine our research objectives. However, beyond this aspect of the analysis, the data collection tool used has, in our opinion, significant cognitive and metacognitive potential. It forces the person who proposes it, as well as the person who participates in the investigation, to take a reflective look at their own thought patterns. The conceptual landscapes obtained provide qualitative and quantitative information on the circulation and construction of knowledge of evolution in the context of its teaching. For these different reasons, the conceptual landscapes, in their final form as well as in their mode of construction, could constitute a tool for teacher training.

8.2. Evolution: greater visibility of epistemological diversity

8.2.1. *A proposal that persists, gradualism that fades*

According to Jean Gayon:

> Evolutionists have become more pragmatic, more pluralistic, and also more contingentist, that is, more attentive to the changing complexity of circumstances that have shaped the history of life (2009, p. 339).

In order to understand this assertion, which nicely summarizes the changes in the evolutionary framework during the 20th century and up to the present day, let us retrace some key stages in the history of this theory, which has obviously undergone transformations and developments since its origin (1859). Notably, the Darwinian proposal of "descent with modifications" has never really been questioned. The same cannot be said about the second pillar, the "theory of natural selection", which has suffered some assaults and lost its hegemonic claim (Gayon 2009; Gayon and Parot 2019).

4. Jörg Zabel and Harald Gropengießer used this term in an article from 2011. We use this term in a different but complementary sense with their kind permission.

The beginning of the 20th century was a low point for Darwinism, but the synthesis achieved from the 1930s onwards between the new genetics and Darwinian principles brought Darwinism back into the saddle (Dobzansky 1937; Huxley 1942; Mayr 1942; Simpson 1944). The main ideas of this "synthetic theory of evolution" (STE) are as follows:

– evolutionary changes are all based on variations that all have a genetic origin;

– natural selection is the main evolutionary mechanism;

– evolution is gradual, i.e. the accumulation of small changes at the molecular level determines the macroscopic changes.

In other words, "evolutionary transformation is adaptive and occurs slowly and gradually, cumulatively, through the addition of small changes under the influence of natural selection" (de Ricqlès and Padian 2009, p. 349).

Let us note the large explanatory power attributed to natural selection with the relegation of genetic drift to an epiphenomenon and, above all, the complete determinism of microevolution (genetic mutations) on macroevolution (the transformation of populations) in a gradual process.

This gradualist approach based on the triptych "mutation, selection, adaptation" would be taught in France from 1988[5] (title of the program: "towards a synthetic theory of evolution"). This date may seem late and, in fact, it is. Prior to this, and since 1945, "theories of evolution" were taught in secondary schools[6]. In reality, this is a question of discussing the respective interests of Darwinism, Lamarckism and mutationism to propose the most relevant evolutionary mechanisms. This significant didactic delay (Quessada and Clément 2007) can be explained in part by a well-documented reluctance[7] by French scientists in the middle of the 20th century to Darwinism. According to de Ricqlès, there were two scientific and "national" traditions that opposed or ignored each other at that time, particularly on epistemological aspects. On the one hand, that of the English-speaking naturalists, led by Darwin, who made room for historicity and contingency, and on the other hand, that of the French biologists, beginning with Louis Pasteur and then Claude Bernard, who were much more focused on the experimental. This could explain why

5. MEN, "Programme de la classe de terminale D", *Bulletin officiel de l'éducation nationale*, 2 juin 1988, supplément au no. 21, p. 49.

6. MEN, "Programme pour l'année 1945–1946 des classes de sciences expérimentales", *Bulletin officiel de l'éducation nationale*, 1 November 1945, no. 5, p. 1900.

7. See the works of Denis Buican, *Histoire de la génétique et de l'évolutionnisme en France* (1984) and Jean-François Picard, *La république des savants* (1990).

even fixism persisted more than would be expected in France, and when it was no longer possible to support it. At that point it was replaced by "a transformism of neo-Lamarckian inspiration" (de Ricqlès 1996).

Meanwhile, in the English-speaking world, from the 1960s onwards and under the impetus of Stephen Jay Gould, the STE that was considered too "adaptationist", i.e. giving excessive explanatory power to natural selection, began to be criticized in the light of the new fields of study and the new knowledge that emerged in the latter part of the 20th century. These views challenge some of the foundations of STE. For example, the development of phylogenetic classification from the 1950s (Hennig 1966) and especially the 1970s[8], Kimura's work on neutral mutations and genetic drift (1968), the discovery of horizontal transfers (Ochiai et al. 1959; Syvanen 1985), the development of research on epigenetics and evolutionary embryology, and the work on punctuated equilibria (Elredge and Gould 1972) and on exaptation (Gould and Vrba 1982), which limit the "creative" power of natural selection.

8.2.2. *Epistemological openness and the diversity of evolutionary problems*

Because the new knowledge that has developed has led to the modification, but not the denial, of the fundamental concepts of the theory, Stephen Jay Gould considers that the Darwinian framework has undergone an expansion with reformulated conceptual contents (Gould 2006). Jean Gayon (2011) speaks of an expansion of the use of concepts and a revision of content. Current topics of discussion include the relative importance of selection and drift, levels of selection and consequences of epigenetics, while some authors discuss whether or not these changes to the evolutionary framework need to be further formalized (Lecointre 2009; de Ricqlès and Padian 2009; Pigliucci 2013; Laland et al. 2014), which, at a minimum, testifies to the importance of the changes taking place.

Yet perhaps the most important change is epistemological. This is partly due to the rise of cladistics, which became dominant in the 1970s after the translation of Willi Hennig's work. The principle is that kinship between living beings is based on the exclusive sharing of homologous characters (from a common ancestor) in a derived state. A character can be in an ancestral state (old form of the character or absence of the character) or a derived state (new form resulting from an evolutionary innovation or presence of a new character) in a group of species studied. For Hennig, only those characters in a derived state and their exclusive sharing are

8. Hennig's book was first published in German (1955) and then in English (1970).

relevant to determining the most related species, which then form a clade or monophyletic group.

This analysis, called cladistics, deals only with the states of homologous characters and their grouping to find the order of common ancestry: "Thus Hennig constructs a system that depends only on genealogy". However, this "order ignores global similarity" (Padian 2004, p. 31). This is a more important shift in approach than it appears. The "evolutionary" or "eclectic" classification associated with synthetic evolutionary theory was also based on a genealogical principle, but it still favored groupings based on global similarities between organisms "rather than those bringing together organisms that are closely related but bear little resemblance to each other" (Padian 2004, p. 30). As a result, in a group of species containing crocodiles, monitor lizards and birds, the crocodile is grouped with the monitor lizard in the "evolutionary" classification, whereas it is grouped with the bird in the cladistic classification. The crocodile shares more characters in the derived state with the bird than with the monitor lizard. The grouping between the monitor lizard and the crocodile is underpinned by a common mode of life, in contrast to the "adaptive leap" represented by flight in birds (Lecointre and Le Guyader 2017, p. 22). Evolutionary processes are therefore determinant in evolutionary classification, and this is the important difference with cladistic classification, which groups crocodile and bird into the same clade. This change in approach gave rise to an important debate between Hennig, father of the original phylogenetic method, and Ernst Mayr, one of the founders of the TSE[9], who undoubtedly understood its profound implications for a position to shake up the "synthetist paradigm" (Gayon and Petit 2018).

This autonomy of anatomical structures (and molecular sequences) with respect to evolutionary processes in cladistic classification has manifested itself in the world of evolutionary research through the more usual use of the terms "process" and "patterns"[10] to describe the objects of evolution. "Processes" are all of the biological mechanisms involved in the transformation and development of the living world (selection, mutation, drift, etc.). "Patterns" are the material and/or historical structures of the living world, present and past (a character, a population, a phenotype, a species, a phylogenetic reconstruction, etc.). They are the result of the chronological summation of the existence of all the temporary actors of evolution (de Ricqlès 2005). This distinction between "patterns" and "processes" was undoubtedly the beginning of a movement to clarify the epistemological status of

9. See, on this subject, issue No. 29 of the journal *Biosystema* (2014): "Analyse cladistique: le débat Mayr-Hennig de 1974", coordinated by Martin S. Fischer and Pascal Tassy.

10. See, in this regard, the titles of the following books: *Phylogenetic, Pattern and the Evolutionary Process*, Olson (1968); *Macroevolution: Pattern and Process*, Stanley, (1980); *Phylogenetic Patterns and the Evolutionary Processes*, Elredge and Cracraft (1980).

evolutionary sciences. According to the proponents of STE, the evolutionary biologist seeks the ultimate causes, i.e. "the causes of the characteristics specific to organisms and in particular of their adaptations" (Mayr 1995, p. 108), which, in the synthetist framework, can only be answered by a selective, gradual and genetic process. The evolutionary "patterns" are thus always dependent on a process of selection and adaptation. It is this link of dependence between "pattern" and "process" that has been loosened today. While evolutionary biology still seeks to reconstruct the unfolding of living forms from their origin and to understand the mechanisms at work, it does so within a more interdependent vision between "patterns" and "processes" (Gould 2002; Gayon 2004; de Ricqlès and Padian 2009; Lecointre et al. 2009). This epistemological openness is, in our opinion, a heuristic framework for thinking about the teaching of evolution. Indeed, it allows us to approach the evolutionary sciences as working in a complementary way on historical problems (the search for evolutionary scenarios) and functionalist problems (the study of the processes at work in evolution). These two types of problems mobilize diversified reasoning and modes of validation.

Concerning the formulation of the updated evolution, we refer to the one proposed by Barberousse:

> Schematically, we can say that the theory of evolution takes as its starting point the fact that organisms can reproduce and that the offspring from these reproductive events can show differences from their parents.
>
> On the other hand, each organism has a limited existence in space and time. The theory of evolution tells us that in this context, the structure of the genealogical network depends on the processes of random sorting (drift) and selective sorting (natural selection) that act on organisms in each generation.
>
> The action of these two processes is determined by both the intrinsic characteristics of the organisms and the spatiotemporal context in which they are found (Barberousse 2011, p. 369).

The first section contains what we will call the conditions of possibility of evolution, i.e. the intrinsic properties of living organisms that allow biological evolution. These are the capacity to vary, to reproduce and to transmit traits to its descendants. Without them, there would be no evolutionary processes as we know them. These three properties of living organisms that constrain evolutionary events are to be mobilized in the work on evolutionary problems and must therefore logically be present in the items collected. The other expected elements are extracted from the following two sections. They are evolutionary processes (selection and drift), patterns (structure of the genealogical network), environmental factors and constraints intrinsic to the organisms. These different elements, structuring the framework of biological evolution, are for us expected elements of the corpus of collected items.

8.3. Construction of conceptual landscapes

Our objective is epistemological and didactic. We seek to identify the structuring elements of the evolution declared by the different audiences interviewed and find out how they are related. The data collection tool and the analysis of the data allow us to access certain characteristics of the frameworks of thought, and the results provide important elements of reflection for thinking about more effective teaching of evolution. The conceptual landscapes we construct are a form of schematization of the representations of groups of individuals, belonging to different communities, on evolutionary concepts and the links that organize them. The groups targeted in this investigation are researchers of evolution[11], teachers of ELS[12] and students in the high school senior science class[13].

Two of the audiences we are targeting are expert audiences: evolutionary researchers and ELS teachers who are trained in the conceptual content of evolution. Both groups are questioned on major scientific concepts of evolution, and we consider that their answers cannot be considered simple opinions or beliefs. We are therefore, in our work, closer to professional representations than to social representations in Abric's broad sense of the term, as the set of opinions, beliefs and attitudes shared and constructed by a group of individuals. This concept was proposed by Moscovici, a specialist in social psychology, in 1961. He also

11. Twenty-six MCF, research engineers, research fellows and research directors; 7 post-doctoral fellows and 16 doctoral students, some of whom belong to the Evolfrance network. Their research units are mainly the following institutes attached to the CNRS: Institut d'écologie et d'environnement (INEE), Institut des sciences biologiques (INSB) and Institut des sciences de la Terre (INSU). The specialties of these researchers are distributed as follows: evolutionary ecology (30%); genetics/genomics (24%); evolutionary biology (18%); modeling/ bio-informatics (13%); phylogeny (8%); developmental biology (5%); paleontology (2%).

12. Sixty-seven ELS teachers: middle school, high school.

13. Seventy-three students from three high school senior classes, from the same high school, with three of the ELS teachers.

developed the concept of professional representations, which is a subgroup of social representations when the subject of the survey is specific to a field of expertise. Moreover, since the students interviewed are not experts, their representations cannot be described as professional.

The resulting landscapes will thus reveal links between the evolutionary concepts for each of the participating groups. This networking of items is important for several reasons. First, it is one of the essential characteristics of their definition, as explained by Guy Rumelhard: "A concept always functions in relation to other theoretical and technical concepts. It is a knot of relations, coherent and organized, and not an element arranged next to others by simple juxtaposition" (Rumelhard 1986, pp. 4 and 13). Second, because these networks, which differ from one group to another, are a representation of the frameworks of thought that we are seeking to bring to light in the form of conceptual landscapes.

8.3.1. *The association network*

The tool used, the association network, comes from sociology and social psychology (Abric 1994, 2003), and is "a study technique for identifying the structure, contents and polarity of the semantic field associated with a certain representation" (De Rosa 2003, p. 81). It is used to access the social representations of a group (more or less homogeneous) concerning an object in the environment under consideration. An association network thus makes it possible to collect items (nouns, adjectives, expressions) based on a stimulus word (a word or expression used to trigger the reflection of the targeted persons), which is the object whose social representations we are trying to discover. When listening to the stimulus word, the participants are asked to first write a list of evocations (spontaneous writing of terms that come to mind in connection with this stimulus word). In the second stage, they were asked to write an importance list in which the items on the evocation list were listed and ranked in order of importance.

The set of items obtained in this importance list is then classified into groups or kernels according to their occurrence and their ranking in the set of responses of the group. The nucleus containing the majority items is called the central nucleus. According to Abric, it has a strong structuring capacity in the sense that it creates consensus and unites the group tested. It is of particular interest to us insofar as it is the result of a reflection based on each person's own knowledge of evolution. It is also the subject of a hierarchy regarding the importance given to the different elements of the scientific theory. It is thus the privileged place of the emergence of representations that we seek to formalize in the conceptual landscapes. To obtain more information on the meaning of the items mobilized and their possible links, we modified the tool. In the original method, the respondents answered the word-stimulus by writing items on a blank page. They can link them by lines, but without specifying the meaning of the links thus marked. As we attach great importance to this networking of items, it seemed more relevant to us to propose that they write lists of items (without linking them), and then ask for a definition, as detailed as possible, of the five items considered most important by each participant. The group's corpus of definitions is then analyzed to identify the relationships between the items in the definitions. This work makes it possible to link together, in a more explicit manner, the main items for each group. This part of the analysis allows us to build a semantic network of items that we then interpret in terms of concepts for two reasons. First, in the importance lists, a very large majority of the items are, not surprisingly, key concepts of evolution. Second, the analysis of the definitions supports the conceptual aspect through the quality and quantity of the links identified between the items. These results are, however, more evident among researchers and teachers than among students, whose definitions are less developed.

8.3.2. *Data collection and processing*

The stimulus word given is: "evolution". The instructions given to the participants are as follows:

– write an *evocation list of* 20 items (noun, adjective, expressions) around a stimulus word given at the beginning of the test;

– then, rank in descending order the items that seem most fundamental from the evocation list. New words may be added. This second list is called the *importance list*;

– write as complete a *definition as* possible for the *first five items* on the *importance list.*

We retained the first five items because in this type of collection the number of hapax (single occurrence item) is often very high and retaining five items allows us to find a satisfactory compromise for our research. The statistical processing requires several steps:

– Harmonization of items (e.g. the items "selection" and "natural selection" will be grouped under the term "natural selection"). This is a critical and delicate step in the sense that we have to standardize while minimizing the loss of information. The definitions are used at this stage to check that a grouping is relevant.

– Determination of the average frequency of each item and its average rank (from 1 to 5) in the importance lists.

– Grouping in four groups of the items collected for the whole group by crossing the average frequency of the items and their average ranks.

	Frequency of item > average frequency of all items	Frequency of item < average frequency of all items
Strong medium rank <= 3	Central core (1)	Contrasting elements (3)
Low average rank > 3	First peripheral core (2)	Second peripheral core (4)

Table 8.1. *Nuclei of the association networks according to the frequency and rank of the items (after Abric)*

According to Abric, the central core has a unifying function that makes it resistant to change. The first peripheral core is also consistent, but more fluctuating in representation and more subject to change. The contrasting elements compile items that are rarely cited, but that have great importance for their authors. The second peripheral nucleus gathers items that are infrequent and placed last in the representation.

The analysis of the definitions of the first five words of each list consists of identifying, for a given item, the possible presence of other items from the importance list in its definition. The items are thus linked together according to their presence in their respective definitions.

We will present the results of each group in two complementary figures:

– The first is a graph representing the items of the central core and the first peripheral core (cores retained as being the most representative), positioned according to their average rank in the importance lists (from 1 to 5) and their average frequency in the responses of the target audience. This first graph is called the "association network" in reference to Abric's nuclei.

– The second is a map that represents the items of the importance lists that are linked together following the analysis of the data from the definitions (when there is a sufficient number of them) and that we call the "conceptual network".

It is the whole of the two figures that we name the "*conceptual landscape*" of evolution. The "landscape" metaphor seems heuristic to us for several reasons:

– The results obtained will bring out the common items of a group of people with regard to the word-stimulus "evolution" and will inform us about their representativeness (and therefore their importance) in this group, but *ultimately* little about their real acceptance by each of the participants.

– The conceptual networks will provide more information, but for a limited number of terms (only the first five from the importance lists). The final result will thus correspond more, in our opinion, to a "landscape" in the sense of a panorama, of an intellectual order, in which the elements of the central core and the peripheral core are organized and delimit the common space of reflection on the evolution of the different participating groups (researchers, ELS teachers and students)[14]. The links established between the items are not named (as in the case of a conceptual map), but remain silent, indicating only a conceptual proximity. The results are initially analyzed as a whole and thus resemble a landscape whose broad outlines are immediately grasped, but whose precise details are not.

– Moreover, the landscapes obtained are not subject to an imperative of scientific validity (as is the case for a conceptual map), each being the result of the analysis of the responses of the different participating groups at a given moment. This specificity makes them informative regarding each group's views on the relationships between the concepts of evolution. In the same way that no two natural landscapes are exactly the same, the results obtained are not superimposable and are not intended to be. Finally, an interesting characteristic of a landscape is its capacity to transform itself. If we can think that the one obtained from the evolutionists is relatively stabilized (susceptible, however, to including new concepts, and also variable according to the disciplines considered), we can suppose that within the framework of a learning process the landscape of the students is brought to be modified. In this case, the initial landscape can become a means of reflection when constructing teaching sessions since it can reveal blind spots or, on the contrary, an excessive focus on certain elements.

8.3.3. *Conceptual landscape analysis grid*

In a second part of the analysis, we contrast constructed landscapes with the categories we began to define above based on a definition of updated evolutionary theory (Barberousse 2011), and its diverse epistemological framework. We conduct

14. The conceptual landscapes constructed by Jörg Zabel and Harald Gropengießer inform us about the cognitive achievements of the students participating in their survey and allow us to visualize the different paths of progression and the recurring obstacles overcome. Their landscape thus represents a learning dynamic, whereas ours is static like a snapshot.

this work with the goal of accessing elements of epistemology that underlie constructed landscapes. In the field of evolution, historical and functionalist problems articulating "process" and "patterns" are worked on. The definition that we take as a reference contains the elements that will be our categories of analysis of the obtained landscapes.

First, we expect the presence of terms relating to the conditions of possibility of evolution, which are the "ontological" properties of living organisms, i.e. variation, the capacity to reproduce and to transmit certain characteristics to its descendants. Evolution is also dependent on the spatio-temporal framework in which it takes place, and we will look for terms that evoke a temporality and those related to the environment. A "process" category will contain the evolutionary mechanisms. The category "pattern" is very vast because it includes the material and historical objects of the living, present and past. On the criterion of the accessibility of an object to the current observation and to its manipulation, we propose to subdivide the "patterns" into two sub-categories: historical and material. The material "patterns" include the current objects that can be observed and manipulated in the laboratory and that are indices of evolution, and the historical "patterns" are past events, unique and inaccessible to observation, such as (e.g. the Alpine orogeny, the Cretaceous-Tertiary crisis, the origin of the first living cells, etc.). In this sense, any current material object (a living being, a fossil, a cell, etc.) is a "pattern" because each object (organic or geological) is the result of history. Past events are also patterns in themselves, but remain inaccessible to observation in their entirety (e.g. a biological crisis, a volcanic eruption). This distinction is important on an epistemological level because, if we think in terms of scientific problems, material patterns, which are in a way the "observables of evolution", can be involved in historical problems and in functionalist problems. However, we have shown in previous works (Paulin et al. 2019) that historical problems (and more generally historical sciences) are not very present in the teaching of evolution in French secondary schools, in contrast to functionalist problems. Moreover, works have long highlighted a tendency towards the experimental in this teaching (Triquet et al. 2012). Differentiating material patterns and historical patterns in the landscapes, especially those of the teachers, can thus bring elements of objectification of these already spotted trends.

On the basis of a theoretical framework, as presented above, we have thus retained the following categories to classify the items of the central and peripheral nuclei: the conditions of possibilities of evolution; the "processes" (selection, genetic drift, mutation, etc.); the historical and material "patterns"; and the elements of the spatio-temporal framework. This analysis grid seems to us to be able to reveal certain epistemological elements of the landscapes obtained and to give them meaning in the didactic aim that interests us.

8.4. Results

8.4.1. *Association networks and conceptual networks*

Below we present the three association networks and the three conceptual networks that we obtained to account for the conceptual landscapes. From the bottom to the top, the items have a higher and higher frequency and from left to the right they are increasingly ranked at the top of the importance lists. For example, among evolutionary researchers, according to the frequency axis, the item "selection" wins, but in the importance ranking it is "change" that is first.

A first difference concerns the number of items mobilized in the central and peripheral nuclei: 16 items for the researchers, 23 for the teachers and 32 for the students. The richness of the corpus is also interesting: it is the percentage of hapax (= a single occurrence) that is always very high in collections of this type, often over 50%. It indicates the diversity of the items used by the respondents. In our case, it is 60% among researchers, 71.4% among teachers and 72.8% among students. We note a gradation from the researchers to the students both for the number of items present in the kernels and for the number of hapaxes, which indicates a semantic dispersion that can certainly be linked to a loss of specialization of the studied scientific field. The items obtained for all of the audiences make the main concepts of evolution broadly emerge. The three graphs of the association networks show a certain visual unity with, in a first approach, a similar distribution that can be described as follows: a few items with a high frequency, including "natural selection", which arrives largely at the top in each of the landscapes, then a second group of items with a lower frequency.

As for the conceptual networks constructed from the analysis of the definitions, they allow us to draw links between the items on the importance lists (according to their reciprocal presence in these definitions). Two parameters are taken into account, one qualitative and the other quantitative. Concerning the qualitative aspect, we look at the homogeneity of the mesh obtained. Can we distinguish groups of items or is it an isotropic mesh? For the quantitative aspect, we count the total number of links that we were able to trace, as well as the number of links per item, which informs us about the density of the network. This linking of items is for us an indicator of the conceptual aspect of knowledge on evolution. We refer in the first place to Ernst Cassirer (1977 [1910]) who insisted on this relational aspect of concepts, which, according to him, designates primarily relations and not isolated objects. In the field of didactics, it is Guy Rumelhard's definition, explained above, that we take up. We think that the analysis of the definitions gives an account of the relational or non-relational aspect of the items cited and therefore of their conceptual nature.

8.4.2. *The conceptual landscape of researchers*

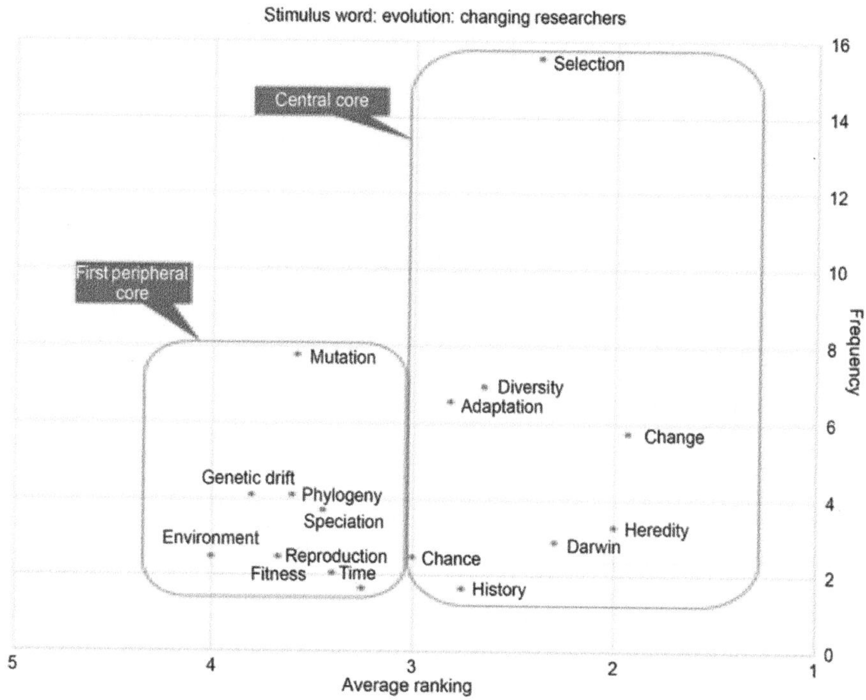

Figure 8.1. *Frequency and rank of items in the researchers' association network*

In the first place, the researchers' landscape (Figures 8.1 and 8.2) is narrowed down to "processes" and mobilizes few "patterns", none of which are material. Evolution thus seems to be thought of primarily in terms of mechanisms, which leads us to a functionalist approach to evolution with the triptych, which we find here, selection-mutation-adaptation at the head of the "processes". Another important point is that the three conditions for the possibility of evolution are present (variation, heredity and reproduction) in a prominent place and make the primordial processes of evolution visible. There are no "material" patterns and some "historical" patterns that relate to the history of life and its reconstruction (phylogeny). Finally, this conceptual landscape of evolution presents a fairly complete panorama of evolution by grouping together the properties of living organisms that condition evolution, the explanatory mechanisms, the spatio-temporal framework and the results of evolution. The characteristic of this network is its absence of material patterns, i.e. observables of evolution, which makes it, finally, more abstract and conceptual than the other networks. The general impression is that the researchers answered as if the questions asked were: "What are the properties that make evolution possible? What are the processes involved?".

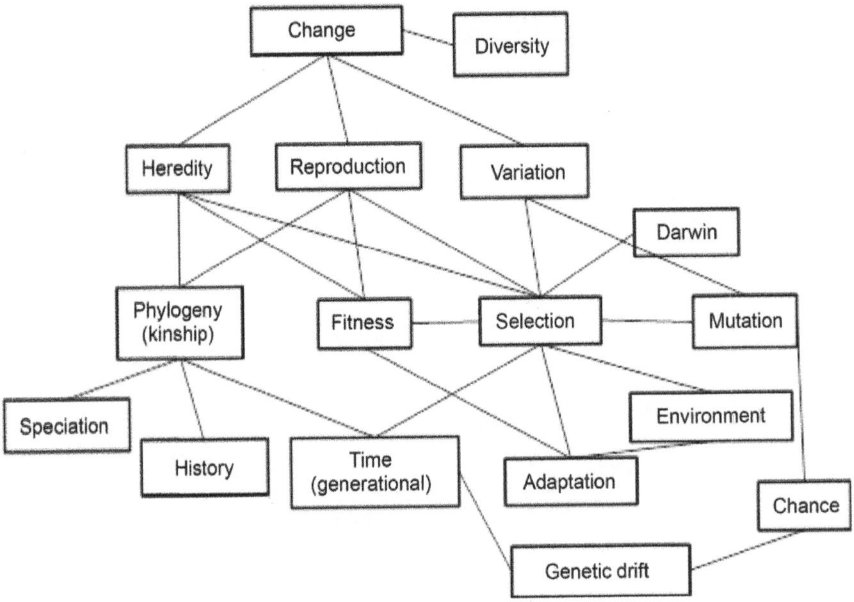

Figure 8.2. *Conceptual network of evolutionary researchers*

8.4.3. *The conceptual landscape of ELS teachers*

Regarding the teachers' conceptual landscape (Figures 8.3 and 8.4), we can see that it is more focused on "material patterns", i.e. the observables of evolution, and especially on the effects (the patterns), than on the reasons or causes of evolution. Teachers, on the contrary, seem to have answered the questions "what are the visible results of evolution?" and "how have species spread out in time and space?". The item "species" is actually well represented here, while it is absent among researchers.

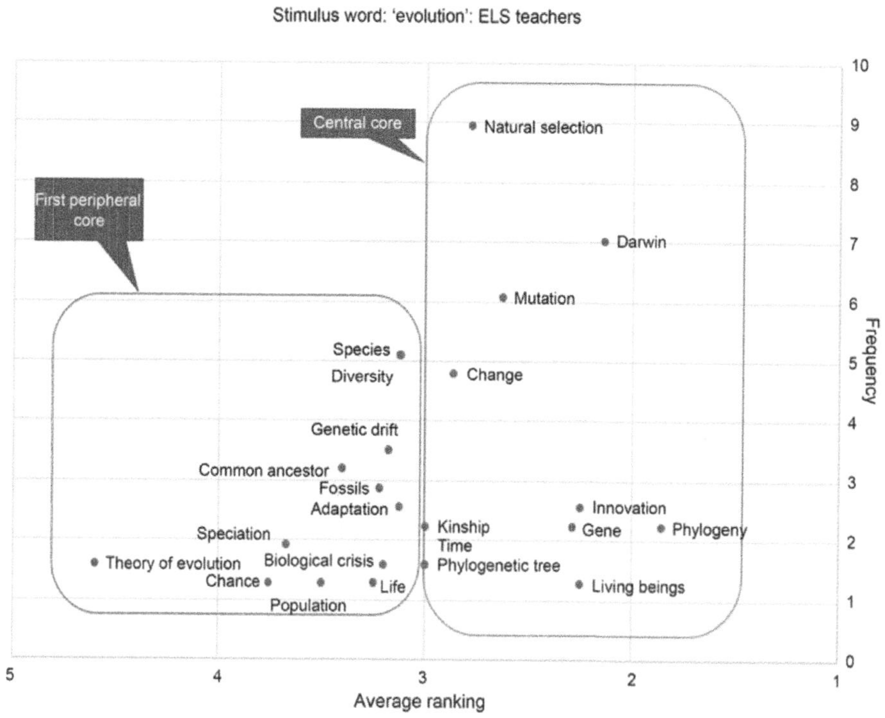

Figure 8.3. *Frequency and rank of teacher association network items*

In comparison, the teachers' networks show more "pattern" than "process" with, on this occasion, material patterns present. The historical patterns that emerge are related to kinship and the phylogenetic classification of the living. However, above all, and this is a real question for us, the conditions of the possibility of evolution are absent. The properties of living organisms that make evolution possible (variation, heredity and reproduction) are not mentioned, and none of the 67 teachers tested used them. Yet these conditions of possibility are indispensable for understanding and explaining the reasons for evolution, and it is these elements that should mark out the space of constraints in which the problems will be worked on. They are, for example, the properties to be evoked to explain the processes of natural selection or genetic drift.

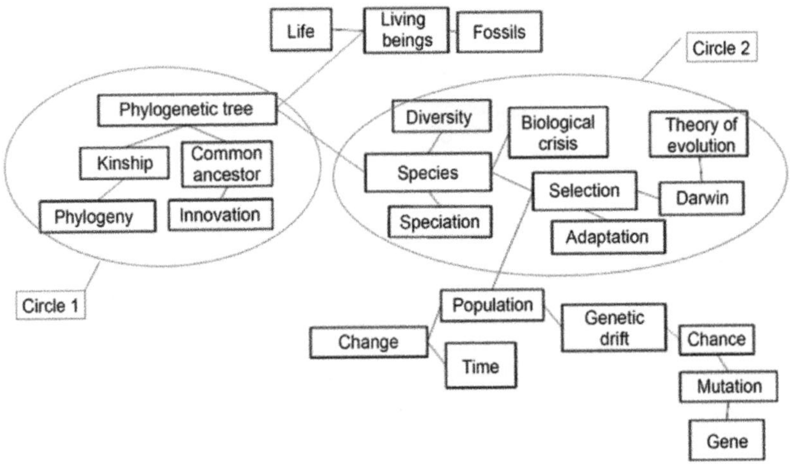

Figure 8.4. *Conceptual network of ELS teachers*

8.4.4. *The conceptual landscape of students in the high school senior class*

What we retain from the students' conceptual landscape (Figures 8.5 and 8.6) is that it brings out the key words of the parts of the program dealing with evolution, which is not surprising. However, it is also rich in material patterns, such as those of the teachers, which feeds the idea of a global representation of evolution focused more on its results and observables than on the processes and their causes. These results are consistent with "the pedagogy of monstration that values anatomical, embryological and molecular facts ..." analyzed by Corinne Fortin (2011, p. 1530) in the teaching of evolution.

The students' landscape presents the greatest number of items, which is not unexpected in view of their non-specialization. The "processes" present partly overlap with those of the researchers and teachers, but we note that an important place is given to the chromosomal and cellular processes of reproduction and fertilization in their genetic aspect. Many patterns are also related to the evolution of human species. The prominence of these two fields seems to us to be related to two chapters of the scientific high school senior class program: the first concerns the diversification of living organisms with the study of meiosis and genetic mixing, and the second is entitled "A look at human evolution" and closes the chapter on evolution. Note, however, the presence of one of the conditions for the possibilities of evolution, reproduction; however, we put this into perspective because it is linked here to fertilization and meiosis and is more in the field of cell biology than the field of evolution.

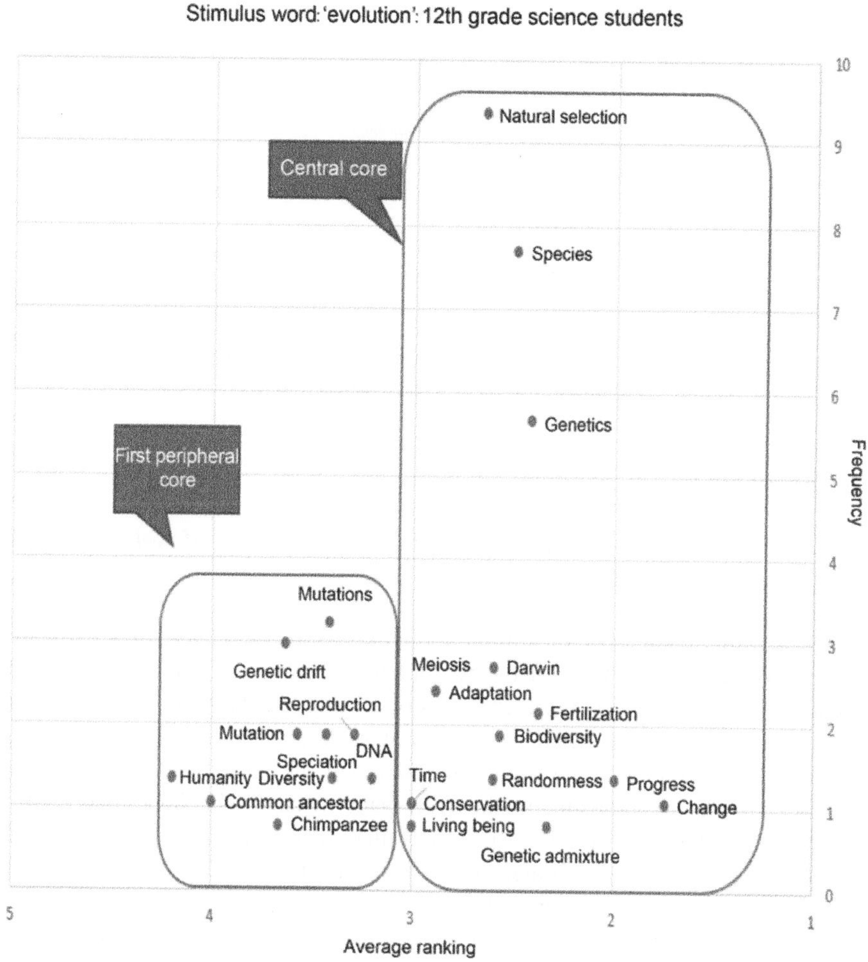

Figure 8.5. *Frequency and rank of student items in the high school senior class*

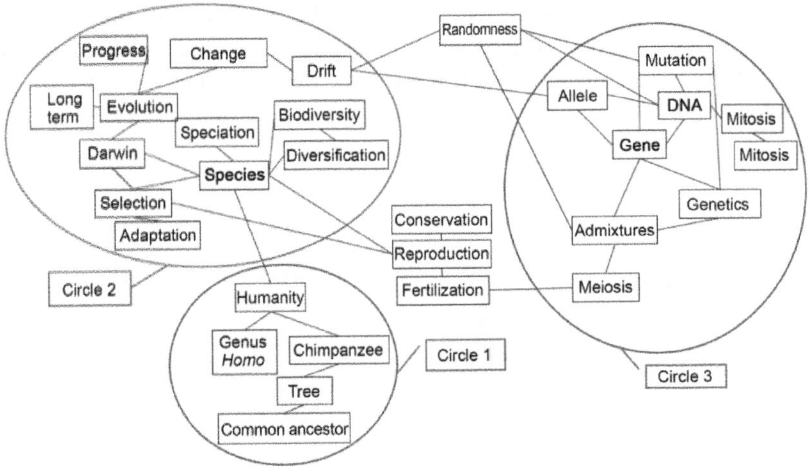

Figure 8.6. *Frequency and rank of student items in the high school senior class*

8.5. Discussion

8.5.1. *Conceptual landscapes: the dominant place of selection and Darwin: the imprint of STE?*

The first important piece of information concerns the item "selection", which is the big winner of the three networks obtained. Selection thus appears to be fundamental in the common representations of the audiences tested. More surprising is its almost systematic association with "Darwin", who theorized this concept. The presence of the author of the origin of species in the evocation lists may seem obvious, but placing him among the five main items of the importance list is somewhat questionable. This is a specificity that "constitutes a rather exceptional situation in the history of contemporary science" (Gayon 2000, p. 205). This hegemonic place of selection immediately brings to mind its status in the synthetic theory of evolution: associated with mutation and adaptation, it was then the explanatory framework of all evolutionary events. We find that these two items are also in a good position in the three networks, especially among researchers: in decreasing order of frequency: selection (1), mutation (2) and adaptation (4). Among teachers and students, "mutation" is in the third position and "adaptation" is placed further back in either the eighth or ninth position.

Another aspect we assume to be an indication of the persistent imprint of STE is the place of genetic drift, which among researchers is in sixth place with a low frequency and comes behind adaptation. It is ranked higher by teachers and students. This may be related to the fact that genetic drift has been addressed in the 10th grade curriculum since 2015 and continues to be taught in the latest curriculum.

Jean Gayon, in an article on evolutionism, recognized that "the vast majority of evolutionists today are still working in the methodological spirit" of the STE (Gayon 1999, p. 463). In 2009, he judged them to be more open to contingency and complexity (Gayon 2009). What is the situation today? Our results, it seems, reveal the strong imprint of STE in the landscapes of evolutionists and teachers. If this is the case, what are the consequences in terms of teaching? Let us recall that one of the obstacles to understanding evolution is the difficult consideration (or even non-acceptance) of contingency in the unfolding of life on Earth. We also know that students are willingly finalists when it comes to explaining evolutionary facts. Teaching evolution within the framework of STE with a predominant selection, leading to systematic adaptation, undoubtedly makes it more complicated to become aware of the historical contingency of evolution and the existence of possible paths other than the ones that have been taken. A more balanced relationship between selection and genetic drift would perhaps guarantee a better consideration of this element of chance in evolutionary events and be better able to counteract the "good-natured" finalism of students.

8.5.2. *Researchers on the side of "processes" and causes, teachers on the side of "patterns" and effects*

This is the most important result of this first analysis for thinking about the teaching of evolution. The comparison of the central and peripheral cores of the three association networks turns out to be rich in teaching and provides some surprises (Table 8.1). It allows us to ask more precise didactic questions about the teaching of evolution. It should be noted that some items do not strictly fit into our categories (e.g. "Darwin") and are therefore not present in the following table, which presents the result of the categorization. In each box, the items are classified in decreasing order of frequency.

	Evolutionists		Teachers		Students	
Conditions of opportunity	Heredity Reproduction Variation				Reproduction	
Process	Selection Mutation Adaptation Genetic drift	Speciation Reproduction Heredity	Selection Mutation Genetic drift	adaptation innovation speciation	Selection Mutation Genetic drift Adaptation Meiosis Speciation	Fertilization Reproduction Chromosomal mixing Mitosis Conservation Diversification
Pattern history	Diversity History Phylogeny		Species Common ancestor Phylogeny Kinship Biological crisis Phylogenetic tree		Species Biodiversity Tree Common ancestor Genus *homo*	
Pattern material			Fossils Genes Living beings		DNA Chimpanzee Humanity Genes	Allele Living being Cell
Frame spatio-temporal	Environment Time		Time		Long term	

Table 8.2. *Comparison of the central and peripheral cores of the three audiences*

8.5.3. *A fragmented conceptual network among teachers, more homogeneous among researchers*

The comparison immediately reveals the fragmented aspect of the teachers' and students' network compared to the more homogeneous (like a spider's web) network of the researchers, which is dense with only a few items, but items that are more strongly connected to each other. The teachers' network is richer in items, but the links are less numerous and organized in islands (black circles on the diagram) that can be moved away from each other. The number of items with only one link is high and accentuates this impression of dispersion of the network. Does this mean that the researchers' knowledge is more conceptual than that of the teachers? Without going so far, this result leads us to question the nature of the items mobilized by the teachers. The curriculum texts regularly use the term "notion" in the objectives to be attained by the students, which in the institutional framework means "school knowledge" and which differs from the epistemological definition of "notion": having a relatively imprecise but generally accepted idea of an object (Duplessis 2007). School knowledge is a progression towards conceptualization that students are supposed to go through during their schooling. As for the teachers, they must master real scientific concepts to be able to accompany this transition. However, the network of teachers obtained in this research indicates this mastery is less complete than expected.

Among the researchers, the conditions of possibility are present and related to the item "selection", which appears as a nodal point (nine links listed). The analysis of the definitions confirms the predominance of this item in the researchers' landscapes. Among teachers, it is the item "species" that is nodal, with five effective links. It is also the item most linked to the others (seven links) among students, followed by the item "DNA" (five links). These two items (species and DNA) are not present among the researchers. Here again, we find the more functionalist aspect of the researchers' landscapes and the more pattern-oriented ones of the teachers and students.

However, if we take a closer look at the islands visible on the teachers' and students' networks (circles 1, 2 and 3), they are the chapters of theme 1A of the high school senior science program, "Genetics and evolution".

We find, in particular, an island on the place of humanity in evolution (circle 1), and another on the evolution of biodiversity and species (circle 2) among teachers and students. Among the latter, we also find an island on genetic aspects and reproduction (circle 3). These results may seem consistent for students who have worked on the different chapters of this theme and mobilize the items in their landscapes according to the division of the programs. This is more surprising for teachers whose overall representation of evolution seems to be more or less based on the key words of the program, but in a fragmented set, according to the chapters studied. This may be the beginning of an answer to our question about the conceptual aspect of the knowledge of the teachers who would have responded with a more academic and programmatic approach to evolution, instead of placing themselves at a more theoretical and global level. Under these conditions, what about the possibility of the students structuring knowledge on evolution if the teachers themselves remain in a fragmented curricular approach?

8.6. Conclusion

In this research, we question the conditions of a teaching of evolution that would lead learners to a problematized knowledge based on a network of concepts that would allow them to overcome the usual obstacles. In our opinion, this objective is based on the teachers' mastery and mobilization of such knowledge in the teaching activities that they will implement. However, the results obtained qualify the reality of this mastery. If we project our results onto the knowledge actually mobilized by the teachers, several points would be problematic:

1) Teachers do not place in their landscape the conditions of possibility of evolution: variation, reproduction and heredity. However, it is on these properties of living things that the theoretical framework of evolution is based. In previous works, we have pointed out the lack of visibility of theoretical frameworks in some students' activities (Paulin et al. 2020), which seems to be confirmed here.

2) The teachers' conceptual landscape is more focused on the effects of evolution than on the causes. For example, their landscape emphasizes diversity or biodiversity, which is a pattern of evolution, and not variation or variability, which is the primary reason. What are the possible consequences of this observation? Problematization requires working on the causes and reasons for events and not just looking for or observing the results. The teachers' landscape also provides elements on the relational and (supposed) synthetic aspect of the teachers' evolution. It appears that the items mobilized are perhaps not scientific concepts in the full sense of the term, which questions the possibility of problematized knowledge in the teaching of evolution.

3) The last observation is that of the strong impact of the program on evolution in the high school senior science class (MEN 2010) and especially of its division into different items that can be found to varying extents in the landscapes of teachers and students. First, this shows, if it were necessary, the meticulous application of the programs dealing with evolution in the classes. However, it also questions the form

of these programs, which seem to strongly determine the structure of the knowledge constructed and which, in the case of evolution, results in a scattering of the objects of this teaching, which perhaps prevents a more conceptual and synthetic knowledge from being achieved. We are touching here on essential curricular questions that require our full attention.

To continue, we would like to emphasize the potential of the tool used in this research and that of the conceptual landscapes that it makes it possible to build. We believe that their mobilization would be relevant in the teachers' curriculum for different reasons.

First, building a conceptual landscape with students allows each one to reflect on their own knowledge, which continues in the phase of prioritizing items and writing definitions. The pooling of each person's answers to the word-stimulus provides a collective perspective that leads to discussions about concepts and, above all, their relationships, which are not common in teacher training. In particular, it can lead to a return to theoretical frameworks and their importance in teaching.

Second, the comparison with landscapes already constituted by an identical or different public leads to an awareness of the many approaches to the theme studied and of representations that sometimes do not totally overlap. It is on the basis of the differences noted that insights on the nature of knowledge and its structuring can develop.

Finally, studying existing or constructed landscapes with students provides information on blind spots, gray areas and excessive focus, and can allow for the preparation of lessons better targeted to the didactic and epistemological difficulties thus identified.

The metacognitive potential of conceptual landscapes and their construction makes this object a powerful learning tool in its capacity to generate sustained discussions on disciplinary, didactic and epistemological levels in relation to the nature and structuring of knowledge. It seems to us that this is what we have the right to expect from teacher training aimed at acquiring the required professional skills.

8.7. References

Abric, J.C. (1994). *Pratiques sociales et représentations.* Presses universitaires de France, Paris.

Abric, J.C. (2003). La recherche du noyau central et de la zone muette des représentations sociales. In *Méthodes d'études des représentations sociales*, Abric, J.C. (ed.). Editions Erès, Toulouse.

Aroua, S., Coquidé, M., Abbes, S. (2002). L'évolution biologique : conceptions et rapport au savoir d'élèves tunisiens. *Actes des XXIVes J.I.E.S*, 265–268.

Barberousse, A. (2011). Pourquoi et comment formaliser la théorie de l'évolution ? In *Les mondes darwiniens – L'évolution de l'évolution*, Heams, T., Huneman, P., Lecointre, G., Silberstein, M. (eds). Editions Matériologiques, Paris.

Barthes, A. and Alpe, Y. (2016). *Utiliser les représentations sociales en éducation.* L'Harmattan, Paris.

Cassirer, E. and Caussat, P. (1977). *Substance et fonction – Éléments pour une théorie du concept.* Editions de Minuit, Paris.

Dobzhansky, T. (1937). *Genetics and the Origin of Species.* Columbia University Press, New York.

Duplessis, P. (2007) La cartographie conceptuelle au service de la didactique de l'information : un outil heuristique pour élucider, enseigner et apprendre les savoirs scolaires de l'Information-documentation. In *Cartographie conceptuelle et didactique de l'information : dix cartes de concepts info-documentaires et étude préliminaire*, Duplessis, P. and Ballarini-Santonocito, I. (eds) [Online]. Available at: http://lestroiscouronnes.esmeree.fr/outils/concept-notion-savoir-scolaire.

Elredge, N. and Gould, S.J. (1972). Ponctuated equilibria: An alternative to phyletic gradualism. In *Models in Paleobiology*, Schopf, T.J.M. (ed.). Cooper and Co, San Francisco, CA.

Fabre, M. and Orange, C. (1997). Construction des problèmes et franchissements d'obstacles. *Aster*, 24, 37–57.

Fortin, C. (2000). Les causes de l'évolution. In *Les formes de causalités dans les sciences de la vie et de la terre*, Rumelhard, G. (ed.). INRP Paris.

Fortin, C. (2009). La métaphore de la parenté est-elle un obstacle à l'idée d'évolution. In *Évolution du vivant, un enseignement semé d'embûches ?* Coquidé, M. and Tirard, S. (eds). Vuibert-Adapt, Paris.

Fortin, C. (2011). L'enseignement de la théorie de l'évolution dans le secondaire : quelques enjeux didactiques. In *Les mondes darwiniens – L'évolution de l'évolution*, Heams, T., Huneman, P., Lecointre, G., Silberstein, M. (eds). Editions Matériologiques, Paris, [Online]. Available at: www.materiologiques.com.

Gayon, J. (2004). De la biologie comme science historique. Sens public, dossier "La représentation du vivant – du cerveau au comportement" [Online]. Available at: http://www.sens-public.org/spip.php?article32.

Gayon, J. (2006). Evolutionnisme. In *Dictionnaire d'histoire et philosophie des sciences*, 4th edition, Lecourt, D. (ed.). Presse Universitaire de France, Paris.

Gayon, J. (2009). Mort ou persistance du darwinisme ? Regard d'un épistémologue. *Comptes Rendus Palevol*, 8(2–3), 321–340.

Gayon, J. (2011). Préface. In *Les mondes darwiniens – L'évolution de l'évolution*, Heams, T., Huneman, P., Lecointre, G., Silberstein, M. (eds). Editions Matériologiques, Paris [Online]. Available at: www.materiologiques.com.

Gayon, J. and Parot, F. (2019). *Darwin et l'après-Darwin*, 2nd edition. Editions Matériologiques, Paris.

Gayon, J. and Petit, V. (2019). *Knowledge of Life Today: Conversations on Biology*. ISTE Ltd, London and Wiley, New York.

Gobert, J. (2014). Processus d'enseignement-apprentissage de raisonnements néodarwiniens en classe de sciences de la Vie et de la Terre. PhD Thesis, Université de Caen Basse-Normandie.

Gould, S.J. (2006). *La structure de la théorie de l'évolution*. Gallimard, Paris.

Gould, S.J. and Vrba, E. (1982). Exaptation – A missing term in the science of form. *Paleobiology*, 8(1), 4–15.

Hennig, W. (1966). *Phylogenetics Systematics*. University of Illinois Press, Urbana, IL.

Huxley, J. (1942). *Evolution: The Modern Synthesis*. George Allen and Unwin, London.

Kampourakis, K. (2020). *Understanding Evolution*. Cambridge University Press, Cambridge.

Laland, K.N., Uller, T., Feldman, M., Sterelny, K., Muller, G.B., Moczek, A., Jablonka, E., Odling-Smee, J., Wray, J.A., Hoekstra, H.E. et al. (2014). Does evolutionary theory need a rethink? *Nature*, 514, 161–164.

Lecointre, G. (2009a). Après la théorie synthétique, quelle biologie ? *TDC Darwin et le darwinisme*, 981, 16–19.

Lecointre, G. (ed.) (2009b). *Guide critique de l'évolution*. Editions Belin, Paris.

Lecointre, G. and Le Guyader, H. (2017). *Classification phylogénétique du vivant*, 4th edition. Editions Belin, Paris.

Lhoste, Y. (2017). *Epistémologie et didactique des SVT – Langage, apprentissage, enseignement des sciences de la vie et de la Terre*. Presse universitaire de Bordeaux, Bordeaux.

Mayr, E. (1942). *Systematics and the Origin of Species from the Viewpoint of a Zoologist*. Columbia University Press, New York.

Mayr, E. (1995). *Histoire de la biologie : diversité, évolution et hérédité*. Fayard, Librairie générale française, Paris.

Munoz, F., Quessada, M.P., Clément, P. (2007). Des analyses statistiques multivariées pour traiter les données issues de questionnaires : conceptions d'enseignants et futurs enseignants de douze pays sur l'évolution. *Actes des Cinquièmes rencontres scientifiques de l'ARDIST*, 297–304.

Ochiai, K.T., Yamanaka, T., Kimura, K., Sawada, O. (1960). Inheritance of drug resistance (and its transfer) between *Shigella* strains and between *Shigella* and *E. coli* strains. *Hihon Iji Shimpor*, 1861, 34.

Olander, C. (2008). The reasoning of students aged 11–16 about biological evolution. In *Biology in Context: Learning and Teaching for the Twenty-First Century*, Hammann, M., Reiss, M., Boulter, C., Tunnicliffe, S.D. (eds). Institute of Education, London.

Orange, C. (2012). *Enseigner les sciences : problèmes, débats et savoirs scientifiques en classe*. De Boeck, Brussels.

Orange, C. and Orange Ravachol, D. (2013). Le concept de représentation en didactique des sciences : sa nécessaire composante épistémologique et ses conséquences. *Recherches en éducation*, 17. Published online 1 October 2013 [Online]. Available at: http://journals.openedition.org/ree/7934 [Accessed 3 July 2021.

Orange Ravachol, D. (2012). *Didactique des sciences de la vie et de la Terre : entre phénomènes et événements*. Presses universitaires de Rennes, Rennes.

Padian, K. (2004). *De Darwin aux dinosaures : essai sur l'idée d'évolution*. Odile Jacob, Paris.

Paulin, F. and Charlat, S. (2020). L'épistémologie des sciences biologiques et géologiques : une occasion d'enseigner l'incertitude. *Raisons éducatives*, 24, 101–126.

Paulin, F., Charlat, S., Triquet, E. (2019). L'enseignement de l'évolution : redonner une place à l'épistémologie des sciences historique. *RDST*, 19, 141–163.

Pigliucci, M. (2013). The nature of evolutionary biology: At the borderlands between historical and experimental science. In *The Philosophy of Biology: A Companion for Educators*, Kampourakis, K. (ed.). Springer, New York.

Quessada, M.P. and Clément, P. (2007). An epistemological approach to French curricula on human origin during the 19th & 20th centuries. *Science & Education*, 16(9–10), 991–1006.

de Ricqlès, A. (1996). *Biologie historique et évolutionnisme. Leçon inaugurale*. Collège de France, Paris.

de Ricqlès, A. (2005). La distinction entre "patterns" et "process" est-elle désuète en systématique ? *Biosystema*, 24, 33–41.

de Ricqlès, A. and Padian, K. (2009). Quelques apports à la théorie de l'évolution, de la "synthèse orthodoxe" à la "super synthèse évo-dévo" 1970–2009 : un point de vue. *Comptes Rendus Palevol*, 8(2–3), 341–364.

de Rosa, A.S. (2003). Le réseau d'association. In *Méthodes d'études des représentations sociales*, Abric, J.C. (ed.). Érès, Toulouse.

Rumelhard, G. (1986). *La génétique et ses représentations dans l'enseignement*. Peter Lang, Bern, Frankfurt am Main, New York.

Simpson, G. (1944). *Tempo and Mode in Evolution*. Columbia University Press, New York.

Syvanen, M. (1985). Cross-species gene transfer: Implications for a new theory of evolution. *Journal of Theoretical Biology*, 112, 333–343.

Triquet E., Gandi M., Guilaud J.-C. (2012). Démarches scientifiques, démarches d'investigation en sciences expérimentales et mathématiques. Evolution des représentations d'enseignants débutants de l'IUFM à l'issue de la formation. In *Démarches d'investigation : références, représentations, pratiques et formation*, Calmettes, B. (ed.). L'Harmattan, Paris.

Zabel, J. and Gropengiesser, H. (2011). Learning progress in evolution theory: Climbing a ladder or roaming a landscape? *Journal of Biological Education*, 45(3), 143–149.

PART 3

Teaching Evolution: Conceptions and Obstacles

This third part proposes new didactic tracks on teaching and learning situations.

The chapter by Denise Orange Ravachol opens this last part by evoking the problems linked to biological crises and the origin of life with a view to scientific acculturation. The aim is to help students overcome catastrophism and common beliefs in order to understand the world today as enlightened and responsible citizens. The author analyzes the conditions needed in teaching to reconstitute a reasoned history of life by looking at the subtle game that evolutionary biology researchers play with time. Yann Lhoste's chapter provides a transition to the following ones as it deals with students' access to population thinking. The author focuses on the difficulties they encounter in mobilizing the concept of species and in working on the scientific classification of living organisms, in the context of primary and secondary school teaching. He shows that the interest in these objects of study seeks to understand how students acquire population thinking. As an extension, in their chapter, Jean-François Poncelet, Christian Orange and Jean-Christophe de Biseau analyze the conditions of construction of the Darwinian concept of natural selection by 11–12-year-old students in French-speaking Belgium. Their study allows us to investigate the problematization capacities of natural selection in young students, much earlier than the current school curriculum offers, and thus provides new perspectives on the problematization of the concept of natural selection in primary school. The chapter by Julie Gobert and Laurent Theis aims to identify the obstacles to the teaching of the ideas of chance in the theory of evolution, within the framework of an interdisciplinary teaching situation of mathematics and Earth and Life Sciences (ELS) in the middle school. The authors seek to advance research on the conditions that could make it possible to learn probabilistic and population evolutionary reasoning, while articulating two disciplines that call upon different epistemologies.

9

Building a Reasoned History of the Living World at School: Under What Conditions?

Denise ORANGE RAVACHOL
CIREL, Université de Lille, France

> *Ainsi durent, ainsi bifurquent l'Univers, la Terre, les vivants et les hommes. Leur avancée ne découle pas d'une source, ne se déduit pas d'un principe, mais n'a cessé d'étonner ceux qui la découvrent et décrivent. Contingente, elle saute et change (Serres 2008, p. 6).*

9.1. Introduction

In France, the Earth and Life Sciences (ELS), recent official instructions have emphasized the appropriation of knowledge and the development of types of reasoning and skills[1] in enabling students to understand the world of today as enlightened and responsible citizens. The exercise of critical thinking is thus particularly emphasized with regard to the lively problems that humanity must deal with, the status attributed to science and its contributions, and the inflationary tendency of information dissemination. In this context, the evolution of living beings and of the planet occupy a central place, scientifically and socially subject to controversy, whether it is a question of admitting them or grasping their mechanisms, of reconstituting them in the past or of projecting themselves on their

1. See, in particular, the Common Base of Knowledge, Skills and Culture (MENESR-DGESCO, 2015, BO No. 17, 23 April 2015, 0 p. 6).

The Evolution of Life: Teaching, Learning and Training,
coordinated by Corinne FORTIN and Julie GOBERT.

future. We are interested in the historical dimensions of these problems of the evolution of living beings. We approach them from an epistemological and didactic point of view, with the aim of establishing reference points for a scientific acculturation of the pupils aimed at freeing them from a catastrophism of common sense and common beliefs to the benefit of the control of modes of reasoning that go beyond this ordinary. Our contribution is focused on the conditions allowing secondary school students to scientifically reconstitute, in whole or in part, a reasoned evolutionary history of living beings. Starting with societal concerns about the history of living beings, we situate our study theoretically, notably as regards the subtle game that researchers in evolutionary biology play with time, and then we focus more particularly on the problem of biological crises and the origin of terrestrial life, two problems that stand out in a global approach to the evolution of living beings, of which there is a trace in the official secondary school instructions[2], and for which students are capable of producing common explanations. We will see that access to scientific historical reconstitutions requires the acquisition of modes of reasoning specific to the historical sciences.

9.2. The history of the living at the heart of societal concerns

As actors in society, students are necessarily impregnated with the problems that run through it in terms of the history of life, especially when it involves the future of humanity. In the current context of "climate disruption" and "finite resources", the history of life and the idea of a sixth extinction are indeed particularly recurrent topics in the news. Legitimate human concerns about the unknown, of which the most fundamental fear remains that of death ("All the partial fears that we name are parts or manifestations of the fundamental fear of death." Delumeau in Delmas-Marty and Delumeau 2013). However, this is also present in paradoxical links between the incitement to behavioral change to stop processes spiraling out of control, at the same time as the idea that a tipping point has already been made and that, in the end, there is nothing we can do about it. And again, in the presentation of contrasting points of view on the future of the living and of humanity. "After humans, the good life, finally", we read in the free daily newspaper *Vingt minutes, Lille-Nord* of October 4, 2019 (p. 14), to say that the disappearance of humans could benefit biodiversity. And also, at the same time, the announcement by the environmentalist

2. For example, in the ELS curricula of the high school senior year science (compulsory education) class of 2001, the ELS curricula of the 9th-grade class of 2008, the ELS curricula of the 10th-grade general studies and technology class of 2019, the science education curricula of the general studies high school senior class of 2019.

Cochet[3] that "the world as we know it will have collapsed by 2030, leading to the sudden death of half the world's population" (*Le Point*, September 27, 2019[4]), in the wake of the publication of his book entitled "Before the collapse – Essay of collapsology"[5] (2019). These uncertain projections of the future (what we have given as examples is indeed only a tiny expression of it) create confusion and worry and stir up quick demands for enlightenment and solutions. At the other end of the time scale of living beings, we find the question of the origins of life, for which the popular infatuation is always sustained (Forterre et al. 2013). In contrast to the question of the fate of the living, this is a relatively "colder" question[6], at least in a number of cultural settings, and one for which many students readily produce an explanation in the form of a small evolutionary history of the living based on an organization of matter at "ground zero" (Orange Ravachol 2012). As Bruner (2002, p. 79) writes: stories "domesticate the unexpected, make it a little more ordinary".

How can we rationally define the history and evolution of living beings, when they become particularly anthropocentric?

9.3. Problematizing to think about a reasoned history of the living

The evolution of living beings is commonly understood as a transformation and a complexification of living forms according to an arrow of time starting shortly after the formation of the Earth up until today. In the course of time, simple microscopic beings (bacteria, etc.) would appear, transform and diversify into larger and more complex living beings. Mammals, to which hominids belong, would represent the most elaborate expression of this in students (Fortin 2009). It is a world where "magical" emergences and omnipotent metamorphoses reign superior. Furthermore,

---

3. Yves Cochet is a French politician and mathematician. He has been Minister of Land Management and Environment and Member of the European Parliament. He was a member of the Les Verts (the Greens) and then of Europe Ecologie Les Verts.

4. https://www.lepoint.fr/societe/il-faudra-s-entraider-ou-s-entretuer-interview-apocalyptique-avec-yves-cochet-27-09-2019-2338175_23.php.

5. Servigne and Stevens introduced the term collapsology in 2015 as "the transdisciplinary exercise of *studying the collapse of our industrial civilization, and what might succeed it" (p. 253).* Their work extends Diamond's (2005) work on the collapse of ancient and contemporary societies.

6. We contrast "cold" questions with "hot", "socio-scientific issues", also called "socially acute" questions. The latter give rise to debate and controversy within the scientific community, in society, and at the level of school knowledge (Legardez and Simonneaux 2011). The question of the origin of terrestrial life is "cooler", in relative terms, for the students in the class we will study next, as their responses show.

it confines us to an Aristotelian vision of progress and hierarchy of beings that is not only extremely prevalent but extremely difficult to combat (Philippe 2013). It is also this model of progress in evolutionary history that Gould talks about, and with which we must break for an inverted vision (maximum anatomical diversity at the beginning of the history of life):

> According to the traditional model of evolutionary history, the "cone of increasing diversity," life continually rises toward greater progress, while expanding outward into more species-from elementary beginnings, with the multicellular animals of the Cambrian, to present levels of progress and diversity" (Gould 1997, pp. 264–265).

In support of this vision of the spatio-temporal deployment of the living, we find the implementation of "magical" durations (it is enough that a long duration is deployed for the most improbable processes to occur), "ad hoc events" (transformations in conformity with what we wish to obtain) and simple causalities (a cause leads to an effect). This is all common sense thinking that cannot satisfy a scientific approach. Indeed, why do we use this temporal sequence and not another? Why do we consider these transformations and their sequence and not others? Finally, why this history of the living and not another? Without working on these questions, in other words, without problematizing, there is a great risk of remaining with a little (storytelling) contingent history (it could be different) (Orange Ravachol 2012).

Scientists have long since broken away from this approach and are constantly working, in an interdisciplinary manner, on reconstructing the evolutionary history of the living. As writes Le Guyader (2019, p. 48):

> Today, evolution is seen instead as a multifactorial process with the astonishing plasticity of the genetic material as its raw material and he points out: the genome "undergoes many more changes over generations than simple point changes" and even "natural selection is no longer the only engine of evolution, as the synthetic theory of evolution postulated (Le Guyader 2019, p .48).

It is clear that:

– On the one hand, while taking into account the Darwinian approach of the 19th century (notably with the principle of descent with modification and the concept of natural selection) or the synthetic theory of evolution (based on two essential ideas: the adaptation of organisms results from the sorting of genetic variation by natural selection; hereditary changes - mutations - are not directed in an adaptive manner,

(Le Guyader 2019, p. 49)), contemporary researchers are now well beyond this thanks, in particular, to the contributions of the cladistic approach in systematics (Hennig 1950[7]), molecular phylogeny, the theory of punctuated equilibria (Elderedge and Gould 1972), genome sequencing and the theory of ontophylogenesis (Kupiec 2012).

– On the other hand, the scientific disciplines contributing to the reconstitution of the history of living organisms strongly articulate the problems of the functioning of living organisms at a given moment and the reconstitutions of their past history. In other words, they explicitly bring into play both synchrony and diachrony. Evolutionary biology is both a functionalist science and a historical science in the sense of Mayr (1989). It cannot therefore fall back on strictly experimental methodologies (repeatable and modifiable tests). It is forced to mobilize modes of reasoning and procedures that account for "the uniqueness of certain factors contributing to an event (…)", factors that "given the laws of probability and the laws of the irreversibility of time, will never again occur together" (Gould 1991, pp. 308–309). Certainly, paleontology and its globalization in making fossils play such a huge role in the prediction and testability of the historical reconstructions they make likely, but by coupling them with phylogenetic reconstructions, they are part of the testing of relative relatedness between taxa (Prochiantz 2010; what we could call with William Whewell the test by additive consilience specific to historical sciences, de Ricqlès 2018), and are linked to cladistics, which, by focusing on the possession of states derived from shared features (synapomorphies), reconstitutes taxon pairings, inferring not only the necessity of common ancestors, but also that of events having affected the genome.

With regard to the common conceptions of the evolution of living beings, linear and progress-oriented, these few epistemological contributions bring us into the scientific complexity of approaches (experimental, historical etc.), reasoning (for reversal of causality, see section 9.4) and figures of time (process time, spatio-temporally singular events). Similarly, they are open to the control of scenarios and processes imagined by scientists. As well as to the tests to refute/validate solutions, but which are of restricted application as soon as someone projects themselves on very ancient times or on processes of very long duration with regard to the human scale, they add a logic of necessities and impossibilities (need for events, kinship). This is the case for the emergence of life on Earth, an example that we will study later. Problematized knowledge with strong argumentative robustness can thereby be constructed: the "knowledge that" the history of living beings would be such is coupled with the

7. For an introduction to Hennig's work, see *Guide critique de l'évolution* (Lecointre 2009).

"why" it cannot be otherwise, to use the expression that Reboul uses for scientific knowledge (Reboul 1992).

9.4. Safeguards and specificity of reasoning in historical problematization

In the paradigmatic framework of contemporary scientists, the history of living beings, which is evolutionary, is rooted in the history of the Earth, which is directionalist[8], and unfolds over the temporalities that exceed human history. This is a difficulty that obliges researchers to provide themselves with safeguards in order to avoid arbitrary and/or overly fixed reconstructions. To this end, scientists mobilize theoretical principles (principles of unity and diversity of the living; Jacob 1997, pp. 170–171) and methodological principles (principle of parsimony, principle of actualism) as indispensable safeguards in the reasoned reconstitution that they make of the world. Let us focus on the principle of actualism (Hooykaas 1970; Orange Ravachol and Beorchia 2011), whose mobilization is essential for several reasons:

– as de Ricqlès (2018) argues, it is impossible to separate paleontology from research on present-day organisms, whether comparative or experimental, because it provides actualist models, data and parameter values, for functional interpretations of past worlds;

– the principle of actualism is:

> the bridge that enables our imagination to transport itself from the present to the past and to evoke, with some confidence in its accuracy, the vision of facts that no human eye has contemplated (Hooykaas 1970, p. 11/1963, p. XII);

– this principle prevents the systematic use of ad hoc events ("catastrophes") to explain the history of the Earth and of life. In doing so, it reduces past events to phenomena identified today.

The reconstitution of the past is thus a matter of a particularly supervised historical problematization. It is a question of co-constructing and articulating in a reasoned manner imagined scenarios and traces (e.g. in the field, traces relating to volcanism, traces of life on the primitive Earth (Westall 2005); in the laboratory, syntheses of prebiotic organic molecules that could prefigure the first forms of terrestrial life), traces that are also part of a genuine conceptual construction. As far

8. The Earth evolves and changes its appearance.

as possible, processes contained in present-day nature are exported into the past, even if it means making necessary the long-time builder of new phenomena (actualism of analogy and long-time actualism; Orange Ravachol 2012). The use of actualism may seem counterintuitive because by exporting present phenomena into the past, and by transforming events into phenomena, this principle seems paradoxically to work against the history it should serve to reconstruct. What we must realize is that it is operative to the point of being carried to its limits, which it reaches for certain problems (the origin of terrestrial life, large-scale biological extinctions such as that which marked the end of the Cretaceous period). It is then that history returns (Tintant 1986). The principle of actualism "does not answer anymore" and the necessity of historical events, singular in time and space, imposes itself, using thought experiments articulating a contingent *unfolding* of history and a *rewinding* stabilizing, in a logic of the necessity of events, of a completed history that is therefore unique (Gould 1991; Orange Ravachol 2012; Orange Ravachol and Doussot 2013): these events are thus not entirely predictable (Gould 1991), and their establishment is a matter of exploring possibilities when following the course of history; on the other hand, rewinding up the course of a completed and thus unique history, they can be established according to a retrodictive propagation of the need for events (Figure 9.1.).

In an interview with Jean-Claude Oliva (October 1, 1997), Gould puts it this way:

> We can say that we are here partly by chance. If we were to rewind the history of life, the probability of the appearance of Man would be very, very low. [...] By using the term "chance", I do not mean to throw [the] dice. In the history of life, it is rather a question of contingency, of the impossibility of predicting what will happen. Once the story has unfolded, we can explain it: in this sense, there is no chance. But we cannot predict it because the possibilities are almost unlimited.[9]

The mark of history is not only those of phenomena still at work today. It can be that of events, which we must think of as temporally and spatially singular constructions, contingent because they might not have happened, and whose salience is recognized because they introduce discontinuities and bifurcations in the course of history. Events are thus identifiable as soon as they become necessary to make a subsequent history possible, which happens when the principle of actualism is carried to its limits. Put differently, events are a kind of residue of actualism (Orange Ravachol 2012).

9. See: https://www.matierevolution.fr/spip.php?article3989.

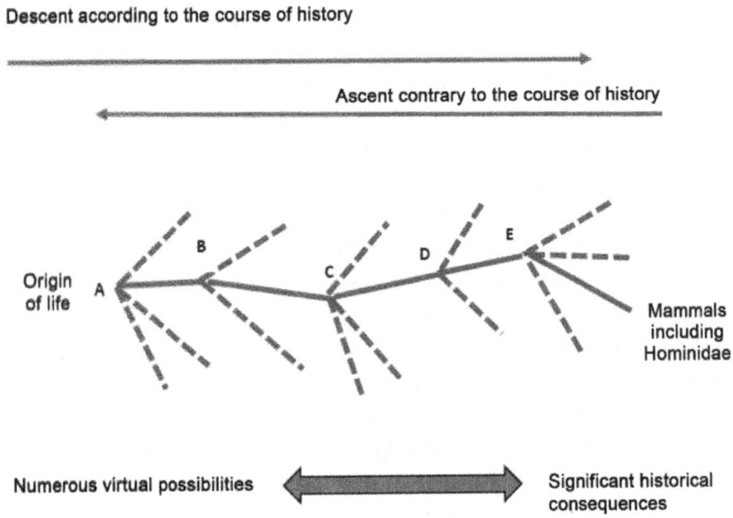

Figure 9.1. *Historical problematization, contingency and event-driven necessities (Adapted from Orange Ravachol 2012)*

The reasoned articulation of the past to the present thus plays on the possibilities, impossibilities and necessities of history. It forces to put into play and to work the contingency, by making other possibilities work. It obliges us to jointly articulate the *rewinding/unwinding* of history with regard to the historical consequences judged qualitatively significant. We thereby show that it differs from common thinking, which is frequently inclined to small stories and oriented by what is true and what is false. In the following sections, we will compare this complex procedure of establishing the need for events with the way students spontaneously reason when it comes to looking back in time. Our choice is biological crises and the emergence of life on Earth.

9.5. Reconstruction of biological crises

Numerous biological crises, massive and rapid extinctions of species and groups mark the history of living beings. The Cretaceous-Tertiary crisis, which secondary school curricula take into account (see, for example, special edition no. 1 of the BO, January 22 2019)[10], is an example that has received a lot of media attention due to the disappearance of most of the dinosaurs. We can consider these crises in two

10. See: https://www.education.gouv.fr/bo/19/Special1/MENE1901647A.htm.

ways: as phenomena, each one then being seen as an example of a biological crisis (functionalist problem); or as events, in their temporal and spatial singularity (historical problem). But Buffetaut writes:

> The only reasonable approach, at present, is to continue the detailed study of each phase of mass extinction - which does not mean, however, that we should abandon more theoretical attempts at explanation. But biological crises are episodes in a history, that of living beings, and explaining one of them does not necessarily explain the others (Buffetaut 1992, p. 48).

The need to treat a mass extinction as a singular event is reinforced if, within the evolutionary framework, it is thought of as a condition of the possibility for the development of mammals, and thus more or less directly for the history of humankind (Orange Ravachol 2012; Orange and Orange Ravachol 2017). This is the case for the near-disappearance of the dinosaurs at the end of the secondary era, which cannot be reduced simply to the mass extinction phenomena identified in other eras. It is important to see that in this context, the "disappearance" event can be seen as a condition of possibility for the development of the mammalian lineage. It is made necessary by the very fact that humankind exists, i.e. in a logic of going back in time (everything happens as if we had an inversion of causality): the existence of humankind implies that almost all of the dinosaurs disappeared. We cannot, however, stick to this "simple" causality, as it would also be the case if we said that the near-disappearance of the dinosaurs led to the emergence of humankind (it is a necessary but not sufficient condition). The reasoning is more complex than a set of "simple" causalities ("this" therefore "that" or "that" therefore "this"). It plays on the possible, the impossible and the necessary. The problematization of such an event is thus a double game (Figure 9.2): an exploration and articulation of the possible (what else could there be?) following the course of history, and a backward propagation of the need for events from a later state. This event thus intervenes in the historical problematization insofar as it is held as a condition of possibility of another event, posterior, whose contingency deserves to be questioned (here the existence of the human species). How then can we think that this contingency makes an event? It is impossible to say that the mass extinction at the end of the secondary era was bound to lead to the development of mammals and the human lineage. However, inversely, with the human lineage having developed, we can retrodictively make the mass extinction at the end of the secondary era a condition of the possibility of this development, and thus a necessity established in an anti-chronological sense. We find the thought experiment proposed by Gould (1991) who speaks of rewinding history and unwinding it again to ask ourselves if we would obtain the same history.

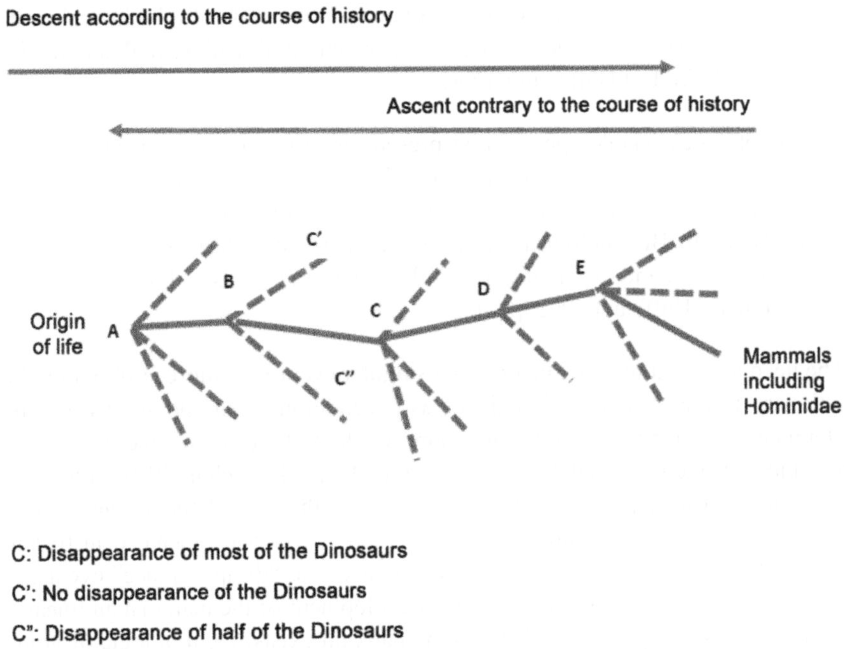

Figure 9.2. *Historical problematization and biological crisis Cretaceous-Tertiary*

This form of reasoning, which couples and articulates the rewinding and unwinding of history with regard to a significant outcome of the latter (e.g. the development of mammals and the human lineage), is out of the ordinary. It is not easy to grasp and mobilize, and we may think that students do not easily appropriate it. It is, however, a prerequisite for the construction of a reasoned history of living beings, and is part of a theoretical and methodological toolkit for critically studying undiscussed and/or simplistic histories of living beings. Moreover, it is interesting to note that it contributes to education for the sustainable development of humanity, of which the school is a strong stakeholder (Orange Ravachol 2020).

We will turn to the students. They are able to spontaneously evoke the biological crisis of the Cretaceous-Tertiary boundary, which they traditionally relate to a spatially and temporally singular catastrophic event. Here is how a student in their final year of science school (17–18 years old) sees it within the history of life. Their words are taken from an interview conducted a few days after they had answered in writing, before teaching and, like the other students in their class, answering a questionnaire on the origins of life. We retain the moments when the exchanges relate to the possibility of new emergences of life during geological times.

> After, after the dinosaurs, I don't know, the dinosaurs disappeared from the surface of the Earth due to a meteorite, yes, that's right, after there was this meteorite, the air must have been difficult to breathe, I don't know if many animals survived, I don't think so, but there may have been an appearance of new bacteria after that. (...)
>
> I thought that, yes, I thought that there were perhaps no animals that had survived and then that well, it had resumed the same sequence and then...
>
> According to what I say, there was a second time, with the disappearance of the dinosaurs, but, yes, so it's certain that even if now, there was, there was a meteorite that, a big meteorite, of a certain diameter that crashed on earth, afterwards, before there were other living beings, it would be, there would be, there would be a certain number of years that would go by before that happened.... because of the air which would be more breathable, thus the impossible life for, that it is the plants or the animals.

This student explains the Cretaceous-Tertiary biological crisis by the fall of a meteorite on the Earth and according to a series mixing in a syncretic way time and causality, we have: (1) the fall of a meteorite that makes (2) the atmosphere unbreathable leading to (3) the disappearance of the dinosaurs and of all animals. Then, after a certain amount of time has passed, he envisages (4) a new emergence of life from simple organisms. To make the extinction massive, he adjusts the size of the meteorite. In terms of modes of reasoning, there is a great difference in comparison to scientists here. The catastrophe to which the student turns refers to disaster and cataclysm (of natural origin here for the case of the Cretaceous-Tertiary crisis; it is not of human origin as for the sixth extinction). It is an event because it is singular in time and space. It involves simple causality (a cause explains an effect). It is attached to the properties of the causes (to a great effect, a great cause). It also brings into play contingency (it can be this cause as much as it can be another one; in this case, it can be the fall of a meteorite as it can be an extraordinary volcanic episode). Contrary to common thinking, scientific thought tries to free itself from this first-level catastrophism and from this ad hoc eventuality, without necessarily discarding them, but in order to construct them in a truly reasoned way, with safeguards such as the principle of actualism.

9.6. The reconstruction of the origin of life

For researchers, the emergence of life on Earth takes on the status of the need for events in the evolutionary framework. It is an emergence that they situate in a narrow temporal window because of the kinship established between present and past living beings and because any idea of "spontaneous generation" is considered to be impossible. This problem is of importance in the reconstitution of a reasoned history of living beings and in the development of the students' critical thinking. Little recognized in its complexity by the official instructions, we know that it does in fact serve the recurrent and virulent creationist attacks in particular, which should not be confined to obscurantism, but in which we should also see militant thinking (Fortin 2009). We are also aware of the tendency of many students to relate it, if not to a "demiurge", then at least to a transformist evolutionary framework where simple microscopic beings (bacteria, etc.) appear and are transformed in the course of time into larger and more complex living beings. In other words, to a world of numerous emergences and metamorphoses, the realization of which is not very constrained.

How do researchers problematize the emergence of life?

The origin of life is a central theme in evolutionary biology (Maurel and Grandcolas 2015). Referring in particular to the work of Reisse (2011), we note that they make it a necessary event in the history of life and that they identify it by articulating a "descendant" approach[11] (from current organisms to LUCA (Last Unicellular Common Ancestor)) and an "ascending" approach[12] (from elementary components to the hypothetical first living system (prebiotic chemistry) (Reisse 2011)). Maurel and Grandcolas speak of a procedure of inference from the comparative study of living things leading to phylogenetic reconstructions, and of exploratory modeling approaches, either purely formal or experimental, which are driven by the reconstruction of processes predicted by the model. As they point out: "reconstructing and modeling are complementary and can stimulate each other" (2015, p. 3). For "modeling approaches can be considered explanatory, and are aimed at provoking understanding or thinking rather than providing an accurate reconstruction in a particular case study" (Ibid, p. 3). Modeling and associated experiments open up the possibility of the emergence of life. In the perspective of the current biological diversity, they invent, control and test processes pertaining to the beginnings of the history of life. They link this work with the contributions of phylogenetic reconstructions that trace back the course of history, and that, by stabilizing some of its stages, close possibilities that could be opened by descending

11. Reisse uses the term "descendant" in the sense of going back to the primordial ancestor, in other words, going back to the beginnings of the history of the living.

12. For Reisse, "ascending" means according to the course of history.

the course of history. We find "on the one hand, the articulation of the functionalist and historical sciences, on the other hand that of the descent/ascent of history characteristic of its reasoned reconstitution".

How do students (*Terminale scientifique* – the last year of specialist science high school (17–18 years old)) deal with the historical problem of the origin of life? And, based on this origin, how do they explain current biological diversity?

Answering a historical question, nothing seems easier on the surface for them: just tell a story (Orange Ravachol 2012; Orange and Orange Ravachol 2017).

Here is a great excerpt from the response of one of them. It is typical of what the student's class was able to produce with the following instruction provided: "At one time, there were no living things on our planet. Can you explain as accurately as possible what happened between the time when there were no living things and the time when the first living things existed on the planet?" and this is what he wrote:

> At the beginning (-4.5 billion years), the Earth was chaos, permanent volcanic eruptions, meteorite bombardments, etc. After the degassing, the atmosphere appeared, because of the oceans life also appeared, but an anaerobic life (primitive soup), using CO2 to live and reject oxygen. After having oxidized all the oxidizable elements, oxygen was fixed in the atmosphere. In the presence of oxygen, aerobic life began, at the bottom of the oceans. This life consisted of bacteria, mollusks, crustaceans, then gradually became more complex, giving birth to fish. These evolved and started to climb on the Earth (yes, fish that walk!). Evolution did the rest of the work, these animals became either saurians that gave birth to dinosaurs or mammals which diversified (the only survivors of a cataclysm which eradicated the dinosaur species!). A species of mammal, the primates evolved to give birth to Man (but that is another story!).

The little stories that the students develop in this way are weakly reasoned. Certainly, they are based on a sequence, marked by "after", "and", "then", the syncretism of time and causality, but it is hardly debatable except by opposing it with another story, without any argument other than its validation by "science". Moreover, all transgressions are possible: they are the ones who introduce complexity and neglect time (Rumelhard 2012).

While such explanations correspond to a fundamental form of common thinking (Bruner 1991), they contrast with the way scientists today think about biological evolution.

– Evolution is not understood as a story that responds to changes in the environment, but as a game of chance and selection in populations of living beings.

– This evolution is not related to ad hoc events (permanent volcanic eruptions… fish that rise to earth, for example). It is thought of in terms of populations and involves current phenomena occurring over a long time (actualism rather than catastrophism).

– Contingency plays an important role: such and such an event took place at such and such a moment, but it could have been another, which means that by rewinding time by thought and unwinding it again, the story could (could therefore have) been another (Gould 1991; Orange Ravachol 2012). One event therefore cannot simply imply another, later one, as in short stories; on the other hand, an established event can make another, earlier one necessary, this one being a condition of possibility for that one.

Let us note that some students, albeit few in number, nevertheless try to think about the origin of life in a retrospective logic starting from the present. This is the case of a student (16–17 years old, repeating the second year of high school), who, during the interview, said several times that she had difficulty imagining the arrival of living beings, as well as that she could not find a beginning. She hears the idea of evolution of living beings (for animals) but is a fixist by force. The following interview extract shows how she reasons.

> E (Student, 20): It's true that it's quite complicated. I try to […] of us and then go back a little bit.
>
> P (Teacher, 21): And what do you find when you go back? Like this.
>
> E (22): Well, I'm having a really hard time finding something, actually.
>
> P (23): But are there […] are some things safer than others? When you're trying to get back up.
>
> E (26): Well, there are things… For me there are always […] plants, yes, I'm sure they've always been […] well […] well, I'm sure […] For me, they have always existed. But the animals, I don't know how… Because there has to be a reproduction but for there to be a reproduction, there has to be […] there has to be… There must always be something in fact!

> You have to […] It's like us, for us to have, for us to have been able to be there, it's that behind it all, there were our parents, all that […] but in the beginning? There was… For us to get here? It's true that it's quite […] I still find people all the time, but at some point, well, there must be something… At some point, it has to stop… There is no beginning. I can't find the beginning, the starting point, how it came about. […] Finally, for me, there was a beginning but […] it's like infinity.

This student constructs the history of the biosphere in a regression from the present, by resorting to her experience as a human being: sexual reproduction where it is impossible to have children without parents; obtaining plants from seeds placed in good conditions, but this retreat does not enlighten her since it places her in front of an impossible choice: to accept an unlimited series plunging into the infinite or to imagine a beginning (Wallon 1945). It can be one or the other. Moreover, if we can envisage a simple beginning of the plants, it is not the same for the animals. Confronted with the contingent and the impossible, we rely on the timeless permanence of life.

The paradigm in which scientists who study evolution are currently situated, with its actualist framework, the long time frame of phenomena, the role of contingency and its anti-chronological necessities, thus strongly constraining the possible explanations – not every event is admissible; none is self-evident – contrary to the way high school students (15–18 years old) explain things in short stories.

On the contrary, for these high school students confronted with a problem of historical biology, the explanatory framework[13] that they mobilize at the outset – explanations taking the form of little stories – does not allow for any work on necessities because it is not demanding enough. To enable them to access reasoned knowledge, they need to adopt an explanatory register that is both closer to that of scientists and more restrictive. Thinking of this learning within the framework of problematization makes visible this didactic condition, which would otherwise remain difficult to identify.

9.7. Conclusion and didactic perspectives

The problems on the history of life that we have just worked on in an evolutionary framework show, through their epistemological support, demanding modes of reasoning that mobilize structuring principles and complex forms of time.

13. In the theoretical framework of learning by problematization, we speak of an explanatory register; Orange Ravachol (2005, 2012).

Because of their strong resonance in the current societal context, we see a certain didactic interest for the training of critical thinking in students. One of the major challenges would therefore be to give them access to "extraordinary" modes of reasoning (Orange and Orange Ravachol 2017) as they have a strong tendency to resort to small stories, *ad hoc* events, circumstantial catastrophism and popular beliefs. In order to construct a reasoned history of the living, what benchmarks for scientific acculturation of students are we able to highlight at the end of this study?

9.7.1. *To free oneself from explanations in short stories (storytelling)*

Confronted with scientific problems (functionalist or historical), the students tell stories and produce ordinary narratives: stake and focus on the characteristics of objects (Piaget's intra-objectal; Piaget and Garcia 1983) whose contingent adventures we follow (they could be different); sequence of episodes where time and causality play in a syncretic way ("therefore/and then"); introduction of ad hoc events. This is a simple, flexible and comfortable way of solving these problems without entering into the phenomenon/event problem and the complexity of explanatory forms of time. However, the students are not fooled, for they also say they are helpless and have trouble imagining what is possible. We must therefore see the historical storytelling they produce as tools (Levi Strauss 1962) to which they have recourse, more out of a need to produce an answer than out of adherence to what they propose (Orange Ravachol 2012, p. 95). Where we find the intelligence of common sense and its narratives that build coherence by assembling the disparate... Where we understand the need to equip students with theoretical and methodological resources to free themselves from it.

9.7.2. *Mobilizing safeguards to avoid first-level catastrophism*

In order to reconstruct a reasoned history of biological or geological objects that go far beyond human time, it is necessary to have means of controlling these reconstructions. This is important, on the one hand, to consolidating the scientific nature of these historical reconstructions, as well as, on the other hand, because these historical problems can sporadically be very prominent in society. This is the case for the evolutionary history of living organisms, against which groups of opponents (creationists, fixist creationists) are particularly influential in society, in a more or less masked manner. The methodological principle of actualism is a powerful safeguard to help students move from small stories to historical explanations, and it needs to be worked on with the students, since its appropriation is not at all obvious. It allows us to construct the necessity of long time as a constructor of phenomena (including evolutionary phenomena) and to free ourselves from "magical" durations and metamorphoses. Carried to its limits, as is the case for the origin of life, and to take contingency into account scientifically, it contributes to the construction of event-driven necessities and bifurcations in the course of history (Orange Ravachol 2012). Note that while its involvement is implicit in the work of contemporary scientists, its involvement is explicitly asserted when it is attacked. The paleontologist Gould, whose tireless struggle on the (pseudo-scientific) terrain of creationism is well known, discussed this principle (1965) while valuing its methodological dimension:

> We should emphasize that the geologist's method of using the present to interpret the past is a particular instance of that inductive mode of reasoning that unites all the empirical science (Gould 1965, p. 921).

It is also to this principle of actualism that the geophysicist Courtillot (1995, p. 240; 2009) refers to in order to give weight, in debates with the "impactists"[14], to his "volcanic" explanation (at the origin of the Deccan Traps) of the Cretaceous-Tertiary crisis. On the one hand, he goes beyond actualism by referring to volcanic phenomena whose magnitude would be unknown today; on the other hand, while the present serves as a key to the past for geologists, it would be possible to think that the past is the key to the future (Ibid 1995, p. 240); a kind of "backwards" actualism (Orange and Orange Ravachol 2017).

9.7.3. *Thinking about the contingency that makes an event by coupling the "rewinding/unwinding" of history*

Like any reasoned reconstitution of history, that of the living involves the construction of the need for events, which forces a problematization of the event/phenomenon couple at the same time as that of contingency. The stakes are high, otherwise reconstitutions will be weakened and ideologies such as intelligent design will be introduced (Fortin 2009, p. 21). The reasoning at work corresponds to thought experiments articulating the unwinding and rewinding of history. It is considered that a past event is a necessary condition for the realization of a subsequent event, a necessary but not sufficient condition, because it leaves room a priori for contingency (the posterior event was not certain at the time of the previous event), but sufficient a posteriori, because the posterior event has occurred and it can only be unique. Prediction (it goes from the causes to the consequences) with consideration toward contingency and retrodiction (it goes from the consequences to the causes) are thus articulated and produce history through the anti-chronological

14. The "impactists" attribute the mass extinction of the end of the Cretaceous to the impact of a large asteroid on the Earth (in the Yucatán, a region of Mexico).

propagation of the need for events. This mode of reasoning, which structures the historical sciences, is not taken into account in ELS curricula. However, its importance can be understood in view of the current focus of teaching on "education for" (sustainable development, health, risks), the anthropization of biological and geological problems, and the opening up of the school to societal concerns about the future of humanity in the context of climate change and attacks on biodiversity.

These are therefore real breaks with common sense thinking that must be considered. Already well documented from an epistemological and didactic point of view, they are a way to develop a well-equipped critical thinking in students.

9.8. References

Bruner, J. (1991). *Car la culture donne forme à l'esprit*. Eshel, Paris (Original edition: (1990). *Acts of Meaning*. Harvard University Press, Cambridge, MA).

Bruner, J. (2002). *Pourquoi nous racontons-nous des histoires?* Retz, Paris (Original edition: (2002). *Making Stories: Law, Literature, Life*. Farrar Straus & Giroux, New York).

Buffetaut, E. (1992). *Grandes extinctions et crises biologiques*. Éditions Mentha, Paris.

Cochet, Y. (2019). *Devant l'effondrement : essai de collapsologie*. Les Liens qui Libèrent, Paris.

Courtillot, V. (1995). *La Vie en catastrophe*. Odile Jacob, Paris.

Courtillot, V. (2009). *Nouveau voyage au centre de la Terre*. Odile Jacob, Paris.

Delmas-Marty, M. and Delumeau, J. (2013). Les peurs, hier, aujourd'hui, demain. Émission Croisements, Extraits choisis. *La lettre n°35 du Collège de France* [Online]. Available at: https://journals.openedition.org/lettre-cdf/2441 [Accessed 8 June 2021].

Diamond, J. (2005). *Collapse: How Societies Choose to Fail or Succeed*. Viking Penguin, New York.

Elderedge, N. and Gould, S.J. (1972). Punctuted equilibra: An alternative to phyletic gradualism. In *Models in Paleobiology*, Schopf, T.J.M. (ed.). Freeman, Cooper & Co., San Francisco, CA.

Forterre, P., d'Hendecourt, L., Malaterre, C., Maurel, M.-C. (2013). *De l'inerte au vivant. Une enquête scientifique et philosophique*. La ville brûle, Paris.

Fortin, C. (2009). *L'évolution à l'école, créationnisme contre darwinisme ?* Armand Colin, Paris.

Gould, S.J. (1965a). Is uniformitarianism necessary? *American Journal of Science*, 263, 223–228.

Gould, S.J. (1965b). Is uniformitarianism necessary? A reply. *American Journal of Science*, 263, 919–921.

Gould, S.J. (1991). *La vie est belle. Les surprises de l'évolution*. Le Seuil, Paris (Original edition: (1989). *Wonderful Life. The Burgess Shale and the Nature of History*. W.W. Norton & Company, New York, London).

Gould, S.J. (1997). *L'éventail du vivant. Le mythe du progrès*. Le Seuil, Paris (Original edition: (1996). *Full House*. Harmony Books, New York).

Gould, S.J. (2006). *La structure de la théorie de l'évolution*. Gallimard, Paris (Original edition: (2002). *The Structure of Evolutionary Theory*. The Belknap Press of Harvard University Press, Cambridge, MA).

Hennig, W. (1950). *Grundzüge einer Theorie der phylogenetischen Systematik*. Deutscher Zentralverlag, Berlin (Revised English translation: (1966). *Phylogenetic Systematics*. University of Illinois Press, Urbana, IL).

Hooykaas, R. (1970). *Continuité et discontinuité en géologie et biologie*. Le Seuil, Paris (2nd English edition: (1963). *The Principle of Uniformity in Geology, Biology and Theology*. Cambridge University Press, Cambridge).

Jacob, F. (1997). *La souris, la mouche et l'homme*. Odile Jacob, Paris.

Kupiec, J.-J. (2012). *L'ontophylogenèse. Evolution des espèces et développement de l'individu*. Quae, Versailles.

Le Guyader, H. (2019). Les mutations de la théorie de l'évolution. *Pour la Science*, 500, 46–57.

Lecointre, G. (ed.) (2009). *Guide critique de l'évolution*. Belin, Paris.

Lévi Strauss, C. (1962). *La pensée sauvage*. Plon, Paris.

Maurel, M.-C. and Grancolas, J.-C. (2015). The quest for the origins in evolutionary biology. *BIO Web of Conferences Proceedings*, 4, 00001 [Online]. Available at: https://www.bioconferences.org/articles/bioconf/abs/2015/01/contents/contents.html [Accessed 8 June 2021].

Mayr, E. (1989). *Histoire de la biologie. Diversité, évolution et hérédité*. Éditions Fayard, Paris (Original edition: (1982). *The Growth of Biological Thought: Diversity, Evolution, and Inheritance*. The Belknap Press of Harvard University Press, Cambridge.

Orange, C. and Orange Ravachol, D. (2017). Problématisations scientifiques fonctionnalistes et historiques en éducation relative à l'environnement et au développement durable : le cas de l'évolution climatique. *La revue des HEP et institutions assimilées de Suisse romande et du Tessin*, 22, 21–38 [Online]. Available at: http://revuedeshep.ch/pdf/22/22-02-Orange-Orange-Ravaloch.pdf [Accessed 8 June 2021].

Orange Ravachol, D. (2005). Problématisation fonctionnaliste et problématisation historique en Sciences de la Terre chez les chercheurs et chez les lycéens. *ASTER*, 40, 177–204 [Online]. Available at: http://ife.ens-lyon.fr/publications/edition-electronique/aster/RA040-08.pdf [Accessed 8 June 2021].

Orange Ravachol, D. (2012). *Didactique des SVT : entre phénomènes et événements*. PUR, Collection Païdeia, Rennes.

Orange Ravachol, D. (2020). L'École au temps des "catastrophes" … Quels repères pour l'EDD ? *Diversité, revue d'actualité et de réflexion sur l'action éducative*, 198, May–August, 102–109.

Orange Ravachol, D. and Beorchia, F. (2011). Principes structurants et construction de savoirs en sciences de la vie et de la Terre. *Education & Didactique*. 5(1), 7–27 [Online]. Available at: https://journals.openedition.org/educationdidactique/1016 [Accessed 8 June 2021].

Orange Ravachol, D. and Doussot, S. (2013). Le catastrophisme en sciences de la Terre et les ruptures en Histoire. In *Les contenus d'enseignement. Approches comparatistes*, Daunay, B., Reuter, Y., Thépaut, A. (eds). Les Presses Universitaires du Septentrion, Villeneuve d'Asq.

Philippe, H. (2013). Origine et diversification des eucaryotes. Conference given on *Hypothèses sur les origines de la vie*, Paris, Institut de France – Académie des sciences, 16 and 17 September [Online]. Available at: https://www.academie-sciences.fr/fr/Colloques-conferences-et-debats-par-et-pour-la-communaute-scientifique/hypotheses-sur-les-origines-de-la-vie.html [Accessed 8 June 2021].

Piaget, J. and Garcia, R. (1983). *Psychogenèse et histoire des sciences*. Flammarion, Paris.

Prochiantz, A. (2010). *Darwin : 200 ans*. Odile Jacob – Collège de France, Paris.

Reboul, O. (1992). *Les valeurs de l'éducation*. PUF, Paris.

Reisse, J. (2011). *La longue histoire de la matière. Une complexité croissante depuis des milliards d'années*. PUF, Paris.

de Ricqlès, A. (2018). *Biologie historique et paléontologie : un regard. Leçon de clôture prononcée le 16 juin 2010*. Collège de France, Paris [Online]. Available at: http://books.openedition.org/cdf/5625 [Accessed 8 June 2021].

Rumelhard, G. (2012). *La biologie, élément d'une culture. Connaître et enseigner le vivant, ... pas si simple !* ADAPT-SNES Éditions, Paris.

Serres, M. (2008). Préface. In *Le grand récit de l'Univers*, Leclercq, B., Jolivet, L., Klein, E., Lachièze-Rey, M., Lehoucq, R. (eds). Éditions Le Pommier, Paris.

Servigne, P. and Stevens, R. (2015). *Comment tout peut s'effondrer, petit manuel de collapsologie à l'usage des générations présentes*. Le Seuil, Paris.

Tintant, H. (1986). La loi et l'événement. Deux aspects complémentaires des Sciences de la Terre. *Bulletin de la Société Géologique de France*, II(1), 185–190.

Wallon, H. (1945). *Les Origines de la pensée chez l'enfant*. PUF, Paris.

Westall, F. (2005). Life on the early earth: A sedimentary view. *Science*, New Series, 308(5720), 366–367 [Online]. Available at: https://www.jstor.org/stable/3841236 [Accessed 8 June 2021].

10

The Concept of Species in Thinking about Evolution and the Scientific Classification of Living Things: Comparative Approach at Different School Levels

Yann LHOSTE
Lab-E3D Université de Bordeaux, Université des Antilles, France, Université Libre de Bruxelles, Belgium

10.1. Introduction

In this chapter, we will focus on the difficulties that students may encounter in mobilizing the concept of species in its population dimension. Indeed, using this concept in class to classify living beings or to study the evolution of species implies, for the students, that one must reason using population approaches that have a transversal and fundamental dimension in biology. Indeed, population thinking is as necessary to understanding food chains in elementary school as it is to thinking about biological evolution in high school and even at university. Indeed, to understand the food chain, we must understand that, in the chain, the rabbit represented is not a specific rabbit (like the one in the back of the classroom), but

The Evolution of Life: Teaching, Learning and Training, coordinated by Corinne FORTIN and Julie GOBERT. © ISTE Ltd 2023.

one that represents the population of all of the rabbits that live in a given place (or that have lived or will live there) and that can be eaten by the foxes that occupy (used to occupy/will occupy) the same environment. Behind the false ease with which the food chain is represented, there is therefore a real challenge for the teaching of biology: that students understand that the living beings represented or studied in the biology course are represented or studied in terms of the populations to which they belong. Access to population thinking is also a condition of possibility for biological evolution. As Jacob reminds us:

> What radically transforms Darwin's attitude towards the living world is the way he considers not individuals, but large populations… The whole theory of evolution is based on the laws of large numbers. Darwin did not resort to complex mathematical treatments to analyze the variation of populations; he was content with intuition and common sense. In considering transformations, he considers only the fluctuations that always occur in large populations, what are called in mathematics the tails of distribution (Jacob 1970, p. 183).

In the same perspective, the scientific classification of life has taken on a new meaning in the framework of the synthetic theory of evolution. As Lecointre reminds us:

> A classification is no longer an art that limits the practitioner to the authority of his or her expertise, but rather a scientific procedure that requires assumptions, formal observations that are transparent to all users, explicit decisions, and above all, exploration of several possibilities (Lecointre 2015, p. 181).

Far from natural history where, for example, Linnaeus' classification sought to "reveal behind the apparent confusion of the world, the sovereign order of nature deriving from divine power" (Duris and Gohau 1997, p. 37), the scientific classification of living organisms seeks to account for the evolutionary history of living organisms through the biological problem of the unity and diversity of living organisms. When we classify species, we seek to answer the following question: which species is closer to which other in terms of evolutionary history?

In view of these issues, work on the scientific classification of living organisms and biological evolution seems to us an interesting object of study in an attempt to understand how students acquire a population-based way of thinking around, in particular, the construction of the concept of species, a concept thematized as an object of discourse within the framework of the scientific classification of living organisms and of biological evolution.

After having returned in a first part to the epistemological and didactic difficulties of the concept of species, we will try to understand how students in the 11th-grade (so, 16 years old) economic and social science stream (ES) and students of kindergarten and 1st-grade (5–6 years old) construct the concept of species before inferring curricular consequences.

10.2. Species as an obstacle to the teaching of biology

10.2.1. *Epistemological approach*

The concept of species, although self-evident in biology, is nevertheless, when examined closely, fraught with difficulty for those interested in classifying living things or explaining evolution, since it does not refer to a univocal definition. As a result, as Gayon (2012) points out in support of the work of Hull (1965), no definition of the category of species is sufficient to cover all the "evolutionary units" that the systematist deals with. Hull concludes by proposing a "disjunctive" definition of the concept of species. This is taken up by Gayon as follows: "Species are populations that:

1) are interfertile;

2) are serially interfertile;

3) meet neither of the first two conditions but diverged from a common ancestor population that met one of these conditions;

4) or do not meet any of the first three conditions but are analogous to populations that meet at least one of the first three conditions" (Gayon 2012, pp. 133–134). Beyond the difficulty of devising a "disjunctive" definition of species and transposing it for classrooms, we can, it seems to us, retain one important point when it comes to teaching biology, which is that a species is a population.

As historians and epistemologists of biology remind us, there is a break in the way of looking at living beings, intervened in the 19th century, that can be considered a condition of possibility for the emergence of the concept of biological evolution and phylogenetic classification. Indeed, there are two ways to consider a series of living beings. Either one is interested in the type to which each living being can be referred. From this point of view, we are not interested in individuals as such.

One acts as if each member of a species were identical to the type; one does not give importance to individual variations. Foucault thus specifies that, for Cuvier, the species constitutes "a reality originally primary and analytically ultimate" (2001, p. 898). For Cuvier, there is an "epistemological threshold" between the individual and the species, a threshold "from which scientific knowledge can begin" (ibid., p. 900). This is what Jacob calls "typological thinking" (1979, p. 148) following the work of Mayr (1976). We are therefore either interested in each living being or in the characteristics that can distinguish it from the type. What matters then is no longer the type, but the population as a whole, through its distribution in space and time (Figure 10.1).

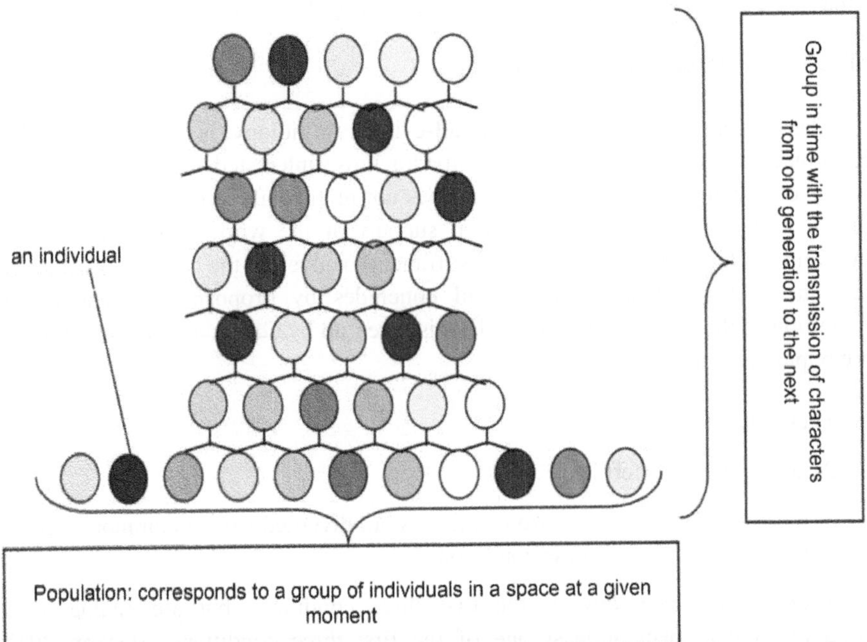

Figure 10.1. *The two sets of living beings for thinking about the concept of species (modified from Lhoste 2008, p. 274; Gobert & Lhoste 2014)*

Two consequences: in this populationist approach, variation is not only a problem of individuals, but of populations. Darwin also shows how "starting from the individual, what can be established as his species, his order, his class will be the rule of genealogy, that is, the succession of individuals" (Foucault 2001, p. 901). The "persistence of the types and plans of organization" is thus no longer "the substratum or the foundation of the history; it is only the consequence of it" (Canguilhem et al. 2003, p. 71). From this point of view, the passage from a typological thought to a populationist thought "marks the beginning of modern scientific thought" (Jacob 1970, p. 191).

When we evoke a rupture in a mode of reasoning, its counterpart, in a Bachelardian epistemology, is the obstacle (Girault and Lhoste 2010). The function of the epistemological obstacle is "to fill the rupture between common knowledge and scientific knowledge"; thus, "it functions in reverse of the No" and "restores the continuity threatened by the progress of scientific knowledge" (Lecourt 1974, p. 27). Epistemological obstacles are not, therefore, difficulties external to knowledge that should be overcome; they have their source in thought itself. Opinions are generally stabilized in depth by obstacles, which implies that during the construction of scientific knowledge "something of the thought has to be destroyed, or at least dismantled in the psychoanalytical mode, so that this one can occur" (Peterfalvi 2001, p. 32).

Thus defined, we can consider that the obstacle to population thinking is essentialism, which Gayon presents to us as follows:

> For Popper, essentialism is a conception of science that consists of explaining phenomena by assertions about "the essential nature of things". For Mayr, essentialism is an erroneous conception of living nature, which neglects or disregards population variability and evolution, and privileges "types" (Gayon 2012, p. 129).

However, concerning the work of scientific classification of living organisms, Gayon notes that "the profound changes that taxonomy has undergone in recent decades, in its methods and concepts, are likely to make us more cautious today. In fields where the fundamental theoretical goal is to classify, even if in an entirely evolutionary framework, it is not obvious that the specter of essentialism is easy to dissolve. For one is quick to renounce morphological essentialism and fall into a 'historical essentialism' (the frantic search for true lineages)" (Gayon 2012, p. 136). In spite of these last concluding remarks by Jean Gayon, we can still consider that the entry into a population thought is a condition of possibility to think about the evolution and the classification of living beings and that essentialism is an epistemological obstacle.

10.2.2. *The concept of species: an objective-obstacle?*

Reflection on the concept of species is a long-standing issue in the didactics of Earth and Life Sciences (ELS) in the French-speaking world, as one of the articles published in the third issue of the journal *Aster: Population* already addressed. They did this by emphasizing that the concept of species would be a clearly identified objective-obstacle (Develay and Ginsburger-Vogel 1987). Works published in the English-speaking world show how essentialist thinking can support an erroneous understanding of biological evolution in relation to the concept of variation (Shtulman 2006; Shtulman and Schulz 2007; Emmons and Kelemen 2015; Alred et al. 2019).

This leads us to the importance of this question in the context of scientific acculturation. To extend the curricular reflections of Astolfi (2008) and Trouvé (2010), it seems to us that we can consider that an element of this acculturation to scientific thought would be to allow students to access reasoning about living beings that are part of population-based approaches.

There is therefore a real didactic challenge in addressing the notion of species both in the context of understanding the evolution of living organisms (Lhoste and Gobert 2009; Gobert and Lhoste 2023) and in the context of classification (Meisel 2010; Halverson et al. 2011; Bosdeveix 2017; Lhoste 2017). Our didactic approach to these questions differs from work conducted in Science Education research in the English-speaking world on similar themes but with different audiences (students, e.g. Meisel 2010; new teachers confronted with the teaching of evolution e.g. Borgerding et al. 2015), different issues (NOS, e.g. Nyléhn and Ødegaard 2018; evolution-religion interaction, e.g. Southcott and Downie 2012) and on different kinds of data and different methods (questionnaires and quantitative approaches, e.g. Alred et al. 2019).

What are the didactic frameworks that we can mobilize to work on this question to analyze the activity of teachers and students in formal learning situations?

10.3. Theoretical and methodological framework

10.3.1. *Modeling teaching and learning situations in terms of structuring contexts*

Let us briefly present the didactic foundations of the approach we have developed on the context structuring model (Lhoste 2017). The scientific acculturation of students is aimed at both the construction of knowledge and the practice of scientific knowledge. It is therefore a question of students constructing ways of acting–speaking–thinking about science. In the dynamics of the teaching–learning processes, we consider that this double simultaneous construction can be described in terms of construction, in the classroom, of a relevant context with regard to the knowledge and the knowledge practices at stake. There is necessarily an epistemological standard in our didactic analyses that we firmly defend. In our modeling, these intentions are concretized in the socio-epistemological context of reference. The activity of the teacher and the students is then looked at from the point of view of the successive and progressive displacements, which take place in the classroom, of the common framework of activity (the context), a context that, at a moment of the work in the classroom, can become relevant with regard to this socio-epistemological reference context. The tools we mobilize to characterize these different contexts are:

– for the socio-epistemological context, the epistemological analyses that were presented above;

– for the problematic and intersubjective context, the objects of discourse and the reasoning that can be inferred from the students' and the teacher's productions in the classroom by relying on tools imported from a pragmatic and enunciative approach to language (Jaubert et al. 2012).

10.3.2. *Problematic and research question*

Knowing the importance of the construction of a population approach to the concept of species from the point of view of acculturation to biology, it seems interesting to us to understand how and to what level of conceptualization students of different ages can access it.

Two case studies will be used to support this research on two different subjects of study: the classification of living organisms in the kindergarten and 1st-grade class (5–6 years old) and evolution in 1st year ES students (16 years old).

In both cases, we seek to understand how students can enter into population-level reasoning. The standard we use for this was presented above, and it is against this standard that we will conduct our analyses. These analyses will focus on two case studies that are contrasted in terms of both the number of sessions involved (one debate session in 1st-year ES versus 14 sessions in kindergarten and 1st-grade) and the degree of openness of the problem situation proposed to the students (a very open question in 11th-grade ES and a more structured situation in kindergarten and 1st-grade).

10.4. First didactic investigation: a case study on evolution in 11th-grade

10.4.1. *The device*

We repeat here a study presented extensively in previous publications (Lhoste 2008; Lhoste and Gobert 2009).

The device was implemented in a class of 11th-grade ES students at Hérouville-Saint-Clair high school by Julie Gobert. During the diagnostic evaluation, the students had to produce an individual text that needed to indicate:

– the causes of the evolution of living species;

– how today's species are no longer those of yesterday.

The students were asked to provide an explanation at the individual and species levels. After this evaluation, the teacher organized group work based on an analysis of the students' individual productions. The groups' productions were used as the basis for a scientific debate.

The debate was conducted by Julie Gobert, filmed and then transcribed. It is from this script, from 110 responses, that we will try to understand how the students mobilize the concept of species to account for the mechanisms of biological evolution.

10.4.2. *Analysis of an extract from the scientific debate*

Table 10.1 presents the didactic analysis of the explanation of the first group of students (see also the poster in Figure 10.2).

These analyses of the interactions at the beginning of the session indicate that the students are trying to construct an object of discourse (what is being talked about: the duck, the species, the duck species) adapted to the proposed situation: the singular (the individual or the type) → plural (a series of individuals: the species?) shifts generate heteroglossia and movements of decontextualization/recontextualization that we have summarized in Table 10.2. It is difficult for the common framework of activity to take shape because of this significant heteroglossia.

1	Teacher	So, we wonder, of course, what can lead in the course of time to the appearance of a new species // what do you think of their explanation / you can give positive or negative criticism or questions	
2	Louise	It's good, it's logical	
3	Lara	It is in relation to reproduction that there are genetic changes / bit by bit as species adapt // finally //	Draft of a reasoning to relate three elements: reproduction – genetic changes – species adaptation. Saying "bit by bit" makes the connection and seems to give adaptation a driving role in the evolutionary process.
4	Teacher	That is to say // the genetic changes take place when in relation to the modifications in the behavior // what precedes // does it happen after or before // you're talking about genetic modifications	
5	Lara	It's when the little duckling comes out of the egg / it happens inside the egg	Decontextualization of a general statement / recontextualization of the specific example of the poster: passage species → individual: "the species" → "the little duckling" → heteroglossia.
6	Jonathan	The genetic transformations happen with the environment //	Decontextualization / recontextualization to a general statement by repeating Lara's reasoning in three/with modifications: - it escapes the question of reproduction; - modification" → "transformations"; - "bit by bit" → "happen with"; - "species adapt" → "with the environment".
7	Jonathan	It is linked to the disappearance of the species: duck 1 disappears in favor of duck 2 //	Rework-modification: "the species" → "duck 1", which is the same heteroglossia as the one identified in 5 and causes a decontextualization/recontextualization movement
8	Teacher	So, duck 1 turns into duck 2 // for Nathan there will be no more individuals of the species duck 1	
9	Quentin	Yes	
10	Louise	If they have undergone the same change in the environment, yes // but if they have not experienced the Ice Age they will remain a duck /	Decontextualization/recontextualization in a more general sense (return of the plural: they, them) by a resumption of the explanation by Louise who clarifies Jonathan's reasoning in 6–7. Same heteroglossia as in 5 and 7: "they will remain a duck".

Table 10.1. *Presentation of group 1: didactic analyses*

2 – Lara	General formulation	
3 – Teacher's question	Remains at general level	
5 – Lara	Decontextualization / general discourse and recontextualization at the level of the case presented on the poster (the duck)	Location-timing of genetic changes
6 – Jonathan	Decontextualization / case study and recontextualization at the level of general discourse	Link environmental modification → genetic transformations
7 – Jonathan	Decontextualization / general discourse and recontextualization on the case presented	Transformation of one species/individual into another
7→10	We remain with the case of the duck	Environmental modification → genetic transformation Long time
12 Léo: "For me the genetic modifications are made at the time of the reproduction / it is to palliate the problems at the level of the cold of the food / I think that there is perhaps a small will of the animal, which inevitably means that there is a reaction after / it is really under the necessity in the face of a climate."	Decontextualization / case study and recontextualization at the level of general discourse	Analogy with learning
14–25: Jonathan in 14: "It's more an adaptation of the body / like us with the human // it's kind of the same thing in relation to tanning or mixing // it's like someone who would live in a hot country / he's going to have his body that will maybe adapt // even for the sun and all that // under the sun he's going to be tanned."	Decontextualization / general discourse and recontextualization on the human case	Mechanism of integration of environmental changes into genetic information Necessity of a long time
27 Nathan: "No, that's why, for example, with insects / we know that they can evolve very quickly // because the renewal rate is / over a human lifetime / it must be several million // so we can see // xxx that's why it works."	Decontextualization / human case and recontextualization on the example of insects	Necessity of a long time
29→ 31	Decontextualization / example of insects and recontextualization on the case of humans	

Table 10.2. *Movements of decontextualization–recontextualization in the first stage of the debate*

These different analyses point to the difficulty encountered by the students in constructing the object of discourse and designating "*who*" is involved in the evolutionary process. This is manifested, when presenting the first group, by heteroglossia in the students' statements, which may use, in the same sentence, the plural and the singular to designate what is involved in evolution (Lara in 3 "***species*** adapt", in 5 "**the little duckling is**"; Jonathan in 7 "it is linked to the disappearance of **the species/duck 1** disappears"; Louise in 10_1 "**they will remain a duck**"). Table 10.2 shows the same type of movement when moving from "duck" to "insects" (since it is easier for a student to identify with a duck than with an insect; in the latter case, the idea of population might be more readily present). It seems to us that, in this study, through the epistemological and language movements identified, the concept of species is put to work. The students encounter difficulties in articulating the horizontal and vertical dimensions needed to construct the concept of species. To justify this, we rely on the singular/plural shifts. In the following parts of the study (which we will not analyze here, see Lhoste 2008), the students reason at the scale of an individual inscribed in a genealogy, which allows the students to take into account a series of individuals over time. The construction of the necessity of a long time (pointed out in Table 10.2) allows them to consider the series of individuals in time. The last intervention, that of Nathan in 27 with the example of insects, could make it possible to begin to link the series in time, which has been discussed above, to the series in space (the population), but his intervention is not worked on. We propose that punctually the context becomes relevant for this student but it is not yet shared by the class.

10.4.3. *Conclusion of this first case study*

On the whole, the students never manage to integrate these two series (which requires them to be situated, to use Figure 10.1, in a plane and not just on lines), yet evolutionary thinking presupposes thinking in the plane. This may explain the difficulties encountered by the students in understanding the concept of evolution. We believe that they do not yet have the biological conceptual tools to think in terms of species (their knowledge of genetics is not related to population genetics, for example).

Even for students who have had an ELS curriculum from grade 6 to grade 10, the concept of species is difficult to mobilize in a relatively open-ended situation and without strong guidance from the teacher. In the remainder of this chapter, we will look at a sequence (as opposed to a debate, as we have just studied) with the aim of understanding how the teacher can support the pupils in constructing the concept of species with pupils in the first grade of high school.

10.5. Second didactic investigation: a case study on the classification of life in kindergarten and 1st-grade

10.5.1. *The device*

Contrary to the sequence studied previously, for this case study, we have a much more substantial corpus (composed of 14 sessions with recovery of individual student productions, as well as collective productions, and the recording and transcription of seven sessions (see Le Marquis 2012)) around the scientific classification of living things. In this chapter, we have chosen to reduce this corpus to excerpts from the beginnings of filmed sessions, which constitute a sort of ritual where the teacher reminds us of the meaning of the work in progress: "what do we want to do in this sequence?"

We hypothesize that the students' answers to this routine question are able to show us, through progressive displacements, how the object of discourse (species) is constructed in the evolution of the understanding/interpretation of the problem situation that structures the sequence, namely classifying 16 animal species (Figure 10.3) around the following question: how can we classify animals to explain that they form families? (The complete device is presented in Figure 10.3.)

10.5.2. *Didactic analyses*

We will follow how students construct the object of discourse during different moments of the sequence. Let us start with the beginning of session three, an excerpt of which is presented below:

1) M: What did we do there?

2) Sylvain: We classified the animals as we thought right.

3) M: Why did we do this?

4) Sylvain: Because there are lots of animals.

5) M: Lots of animals that are what?

6) Cassandra: Not the same.

7) Aurélien: Different.

8) M: And we're trying to see…

9) Daphné: If, for example, I put the jungle with the jungle. I put those who slide with those who slide.

10) M: What are we trying to see, what are we trying to do?

11) Guilhem: To classify them in the right family.

The use of the term "animals" at the beginning of the session refers to an interpretation based on the everyday concept of the problem situation (each vignette represents an animal, an individual that one can name). We can refer these first formulations to a typological conception of the species where the name of an animal refers to a prototypical representation of the species (a rabbit: four legs, a tail, long ears).

Here is an excerpt from the beginning of session five:

1) M: What are we doing? Why are we doing it? What have we already learned?

2) Sylvain: We practice sorting animals to put them in the right family.

3) M: Why do we **classify animals** to put them in the right families?

4) Daphné: Because, we saw, well, in fact we saw that, what the animals have, we said that now we were looking at what the animals had because they always have them.

5) M: The others? Is that why we try to put them in the right families? Because we look at what the animals have on them?

6) Cassandra: We **put them in the right family** because… we…

7) M: Help her then!

8) Sylvain: Because there are many species of animals.

9) M: And what are these **species of animals like**?

10) Antoine: They don't look alike.

11) M: They are…

12) Class: Different.

13) M: But what did we see when we looked at all these species? What did we notice?

14) Sylvain: There are **lots of them**.

15) M: There are lots of them. We saw that there were many different ones.

16) Lou-Anne: We've seen that there are over 100 million.

17) M: We saw that there were several million species, yes. But what did we see when we look at all these species? Daphné said it a little bit earlier. What did we notice?

18) Lou-Anne: What they have on them.

19) M: Why do we look at what they have on them?

20) Aurélien: Because they always have it.

21) M: Because they always have that. We're sure of what they have. But why are we looking at what they have on them? What are we trying to do?

22) Mathys: To sort them out.

23) M: We're trying to get them to…

24) Sylvain: **Classify**.

25) M: We're trying to classify them. Right. So, this work, here you did it a second time last week. I gave you **all the animals** again, all the 16 vignettes and I asked you **to classify them** and as Daphné said, I asked you to look at what **to classify them**.

As early as session five, we can see the appearance of the term "species", which at first seems to be used in the same way as "animals", always in a typological conception of the species. We note that it is the term "species" (sessions seven, nine) that is used by the pupils alone or in mixed formulations such as saying "species of animals" progressively during the sequence (session 10).

The formulations in session 11 are particularly interesting to consider as there is explicit negotiation around the meanings of the terms "animal", "animals" and "species" (in bold in the excerpt below). Here is an excerpt from the beginning of session 11:

130) M: So, to classify the animals, you have to find something that works for –

131) Sullivan: All of them.

132) M: All of what?

133) Daphné: **All cows**.

134) M: All cows, all frogs,

135) Jules: **All snakes**.

136) ?: All bears.

137) M: So, what's it called?

138) Sylvain: **The species**.

139) M: All animals of the same –

140) A few people: Family

141) Sylvain: **Species**.

In 130, the teacher takes up the formulation of the problem situation to "classify the animals", which is immediately the subject of a discussion: "all" → "all of what" → "all the /animal names/" → "the species" → "all the animals of the same family" → "of the same species". This succession of reformulations attests to a rise in generality from concrete examples of animals to the notion of species, which includes all animals that resemble each other (i.e. that have the same hereditary characteristics from a biological point of view). Where initially the discourse object "species" referred to a typological conception of the species, these negotiations lead the class to conceive the species in a population approach "species – all animals of the same family". We can say that the intersubjective context starts to become relevant as we see that the students begin to distance themselves from typological thinking.

This schematization of the concept of species is repeated a little later in this session (in bold in the corpus below).

Extract from the continuation of session 11:

243) M: Yes, that's what Guilhem was telling us earlier. And we had seen that the embryo of the frog had a tail. The tadpole, you remember? And that it disappears afterwards. Have we done the trick or have we learned something else to classify the **animals**? Important to know how to do it and to succeed in the classification? I see that here we haven't written anything. Look at it. We have put that it is necessary to look at **what the whole species has** and we have not put anything in front of it. Why is it necessary to look at **what the whole species has**? This is perhaps a little difficult!

244) Sylvain: It's because otherwise we're only talking about one animal and not the whole species.

245) M: Yes. We're not talking about **an animal**, we're actually talking –

246) Jules: **Animals**.

247) M: **From a** –

248) A few people: **Species**.

249) M: Species. When we make a classification of animals, we do not classify a –

250) Antoine: Alone.

251) M: One animal. We classify –

252) Lou-Anne: Animals.

253) M: We file, no!

254) Sylvain: **Species**.

255) M: What do we classify? Do we classify an animal or do we classify **animals of the same** –

256) Lou-Anne: **Species.**

257) A few people: **Species.**

258) M: You don't classify one animal, you classify the whole species.

259) Adèle: And that's it.

260) Sarah: We're all done!

261) M: (Writes) Because we don't file one, one what?

262) Daphnè: An animal.

263) M: What are we filing?

264) Hugo: **Species.**

The tasks proposed by the teacher in the work prior to this session (work on the embryos of frogs and boas) made it possible to return to the meaning that the class was able to attribute to the objects of discourse ("animal", "species"). When we sought to establish a scientific classification of living things, it was a question of constructing "families" in which the individuals that make up the family have the same characteristics: Teacher: "Why do we have to look at what the whole species has?" → Student: "It's because otherwise we're only talking about one animal and not the whole species". This sequence indicates that Sylvain constructed the "species"/"animal" distinction that we commented on above. A species groups together all the individuals that share the same characteristics, even though each individual may have its own characteristics (all cows have four legs and a tail, even though they may have different colored coats), which led the students to specify that it was not an animal that was being classified, but one species or multiple species. It seems to us that this construction of the discourse object "species" indicates that the students at this point in the sequence were mobilizing a population approach to the species, which is relevant to understanding the scientific classification of living things. With regard to the species worked on in this sequence, the population approach is highly limited, as shown in Figure 10.1, to a series of individuals at a given time.

10.5.3. *Conclusion of this second case study*

With regard to the activity of scientifically classifying living organisms, it seems to us that we can conclude from these analyses that we have an indicator (species in a population conception) that attests that the intersubjective context has become relevant with regard to the knowledge in play. It seems to us that we can link the relevance of the intersubjective context with regard to the knowledge at stake to the different activities proposed to the students, as well as to the recurrent organization over a long period of time (14 sessions) of the devolution to the students of the biological meaning of the activity in which the teacher wishes to engage them.

10.6. Discussion and conclusion

The construction of the concept of species in a population approach seems to us to be an important issue in the teaching of biology and in the acculturation to this branch of science. Our case studies show the difficulty students of different ages have in distancing themselves from a typological conception of species in order to access reasoning at the population level. In the case of classification in kindergarten and the 1st-grade, we have highlighted the need for a long time[1] to consider the species as a series of individuals at a given time that simultaneously share common attributes and certain variable characteristics (such as the color of the cow's coat). It should be pointed out that in order for the long term to allow the students to move away from the typological approach to the species, the teacher must work hard to develop the same problem situation, the backbone of the sequence, a problem situation whose meaning is renegotiated at each session and that ends up becoming meaningful for the students. In the case of the session studied in the first year of secondary school, the short time of the scientific debate based on a very open situation only allows some students to access reasoning at the population level. Even if the object of studying evolution is more conducive to relating the two series of individuals that constitute the species, as identified during the epistemological analysis, we can only note that, over the time frame of one or two sessions, it seems difficult to allow the students to conquer a population thinking.

This set of results is in line with others presented elsewhere (Gobert and Lhoste forthcoming), where this issue of including the same family of problem situations in the long run of a class sequence in 3rd-grade, with significant support from the teacher, enabled the students to conquer this population thinking.

---

1. Not in the sense of a long time in the history of the Earth or of life, but of a long time when it comes to working in class.

If it is therefore possible for very young children (5–6 years old) or those a little older (those in 3rd-grade, so 14 years old) to enter into reasoning by mobilizing a student's population conception. It should be noted that this is, in our case studies, the fruit of work that must be done over time, that must be strongly supported by the teacher and that is never finished and must always be taken up again, as shown by the difficulties encountered here by older students (16 years old). It seems to us that this perfectly characterizes the obstacle of typological thinking, which falls under the bracket of common sense thinking and a certain form of essentialism (Shtulman 2006; Shtulman and Schulz 2007; Emmons and Kelemen 2015; Alred et al. 2019); an obstacle that should be regularly worked on by teachers in biology courses at all levels of schooling and even in the context of teacher training (see Bosdeveix's (2016) work in ELS teacher training).

10.7. Appendix

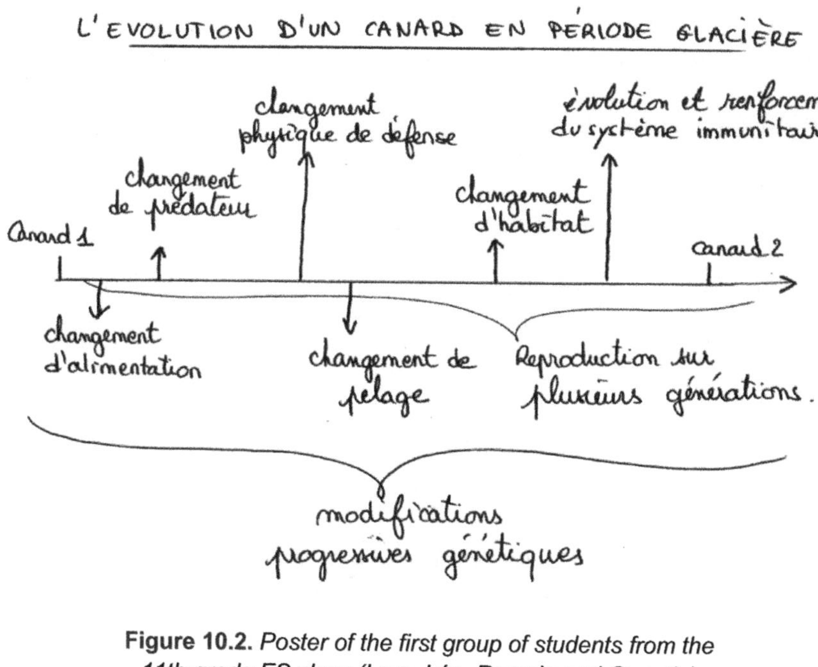

Figure 10.2. *Poster of the first group of students from the 11th-grade ES class (Lara, Léo, Romain and Quentin)*

Finalization announced to the students: explain how the animals around us are classified
Goal of the task: classify the animal images
Starting question: how can we classify animals to explain that they form families?
Constraints:
– Final classification must be common to the whole class
– One animal is only represented by one picture
– There must be at least two pictures per group
– It is necessary to name the groups created (give the possibility of having a group called "I don't know")
Success criteria:
You are successful if each species can be found
You are successful if you know how to explain for each species how it has been placed in that particular family

Figure 10.3. *The situation-problem structuring the classification project in kindergarten and in the 1st-grade*

10.8. References

Alred, A.R., Doherty, J.H., Hartley, L.M., Harris, C.B., Dauer, J.M. (2019). Exploring student ideas about biological variation. *International Journal of Science Education*, 41(12), 1682–1700.

Astolfi, J.-P. (2008). *La saveur des savoirs. Disciplines et plaisir d'apprendre*. ESF Éditeur, Issy-les-Moulineaux.

Borgerding, L.A., Klein, V.A., Ghosh, R., Eibel, A. (2015). Student teachers' approaches to teaching biological evolution. *Journal of Science Teacher Education*, 26(4), 371–392 [Online]. Available at: https://doi.org/10.1007/s10972-015-9428-1.

Bosdeveix, R. (2016). Entre classification fonctionnelle et phylogénétique : le groupe des végétaux. Une reconstruction didactique fondée sur l'histoire des sciences dans le cadre de la formation des enseignants de sciences de la vie et de la Terre. PhD Thesis, Université Paris-Diderot-Paris 7.

Bosdeveix, R. (2017). Les raisonnements classificatoires de futurs enseignants de SVT sur le groupe des végétaux. *Recherches en didactique des sciences et des technologies*, 16, 57–92.

Canguilhem, G., Lapassade, G., Piquemal, J., Ulmann, J. (2003). *Du développement à l'évolution au XIXe siècle*. PUF, Paris.

Develay, M. and Ginsburger-Vogel, Y. (1987). Population. *Aster*, 3.

Duris, P. and Gohau, G. (1997). *Histoire des sciences de la vie*. Nathan, Paris.

Emmons, N.A. and Kelemen, D.A. (2015). Young children's acceptance of within-species variation: Implications for essentialism and teaching evolution. *Journal of Experimental Child Psychology*, 139, 148–160.

Foucault, M. (2001). *Dits et écrits I. 1954–1975*. Gallimard, Paris.

Gayon, J. (2012). De Popper à la biologie de l'évolution : la question de l'essentialisme. *Philonsorbonne*, 6, 127–136.

Girault, Y. and Lhoste, Y. (2010). Opinions et savoirs : positionnements épistémologiques et questions didactiques. *Recherches en didactiques des sciences et des technologies*, 1, 29–66.

Gobert, J. and Lhoste, Y. (2023). Conquérir un point de vue populationnel et historique pour apprendre l'évolution. Analyse de la co-activité professeur-élève par l'étude de la structuration des contextes. *Éducation & didactique*, 17(1).

Halverson, K., Pires, C.J., Abell, S.K. (2011). Exploring the complexity of tree thinking expertise in an undergraduate systematics course. *Science Education*, 95(5), 794–823.

Jacob, F. (1970). *La logique du vivant : une histoire de l'hérédité*. Gallimard, Paris.

Jacob, F. (1979). L'évolution sans projet. In *Le darwinisme aujourd'hui*, Chapeville, F. (ed.). Le Seuil, Paris.

Jaubert, M., Rebière, M., Bernié, J.-P. (2012). Communautés discursives disciplinaires scolaires et construction de savoirn : l'hypothèse énonciative. *Forumlecture.ch*, 3.

Le Marquis, H. (2012). Caractérisation du vivant au cycle 2 et cadre phylogénétique actuel : quel projet d'enseignement pour construire et maintenir le sens en prenant en compte les logiques des élèves, du savoir et de la société ? PhD Thesis, Université de Caen-Basse-Normandie, IUFM.

Lecointre, G. (2015). Descent (filiation). In *Handbook of Evolutionary Thinking in the Sciences*, Heams, T., Huneman, P., Lecointre, G., Silberstein, M. (eds). Springer, Dordrecht, Notion.

Lecourt, D. (1974). *Pour une critique de l'épistémologie (Bachelard, Canguilhem, Foucault)*. François Maspéro, Paris.

Lhoste, Y. (2008). Problématisation, activités langagières et apprentissages dans les sciences de la vie. Études de débats scientifiques dans la classe dans deux domaines biologiques : nutrition et évolution. PhD Thesis, Université de Nantes.

Lhoste, Y. (2017). *Épistémologie & didactique des SVT. Langage, apprentissage, enseignement des sciences de la vie et de la Terre*. Presses universitaires de Bordeaux, Pessac.

Lhoste, Y. and Gobert, J. (2009). Le concept d'espèce, outil/obstacle pour comprendre l'évolution biologique : le cas d'une problématisation scolaire en première ES. *Actes des sixièmes rencontres scientifiques de l'ARDiST. Nantes*, 14–16 October.

Mayr, E. (ed.) (1976). Typological versus population thinking. In *Evolution and the Diversity of Life. Selected Essays*. Harvard University Press, Cambridge, MA.

Meisel, R.P. (2010). Teaching tree-thinking to undergraduate biology students. *Evolution: Education and Outreach*, 3(4), 621–628 [Online]. Available at: https://doi.org/10.1007/s12052-010-0254-9.

Nyléhn, J. and Ødegaard, M. (2018). The "Species" concept as a gateway to nature of science. *Science & Education*, 27(7), 685–714.

Peterfalvi, B. (2001). Obstacles à la compréhension des transformations de la matière. Obstacles et situations didactiques en sciences : processus intellectuels et confrontations. L'exemple des transformations de la matière. Thesis in science and education, 179–226. Université de Rouen.

Shtulman, A. (2006). Qualitative differences between naïve and scientific theories of evolution. *Cognitive Psychology*, 52(2), 170–194.

Shtulman, A. and Schulz, L. (2008). The relation between essentialist beliefs and evolutionary reasoning. *Cognitive Science: A Multidisciplinary Journal*, 32(6), 1049–1062.

Southcott, R. and Downie, J.R. (2012). Evolution and religion: Attitudes of Scottish bioscience students to the teaching of evolutionary biology. *Evolution: Education and Outreach*, 5(2), 301–311 [Online]. Available at: https://doi.org/10.1007/s12052-012-0419-9.

Trouvé, A. (2010). *Penser l'élémentaire. La fin du savoir élémentaire à l'école*. L'Harmattan, Paris.

11

Conditions for the Construction of the Darwinian Concept of Natural Selection by 6th-Grade Pupils in French-Speaking Belgium

Jean-François PONCELET, Christian ORANGE and Jean-Christophe DE BISEAU
Université Libre de Bruxelles, Brussels, Belgium

11.1. Introduction

Evolution is a complex teaching subject that is approached in French-speaking Belgium through three complementary entry points: classification, evolutionary facts and theories. Although it occupies a central place in modern biology, it is only addressed at the end of the high school curriculum. The classification of living things must be partially learned at the end of the first year of middle school (13–14 years old), while facts and theories must be addressed during the last three years of high school (16–18 years old).

This approach has already been questioned by many didacticians who suggest starting to teach evolutionary theory earlier (Jimenez-Aleixandre 1992; Desmastes et al. 1995) because it is a complex subject. Our work is in line with research questioning and documenting this proposition. In particular, we are interested in teaching natural selection as an essential explanatory mechanism of species change over time. Following the existence of several studies on its teaching in the middle of the high school (Gobert 2014; Coupaud 2015), we decided to investigate the abilities of students to take charge of a problematization about natural selection in a 6th-grade class (end of elementary school in Belgium, 11–12 years old) in French-speaking Belgium.

Learning about natural selection does not appear in the official texts that specify the skills to be acquired at the end of elementary school, even though they authorize teachers to make elementary school students aware of themes into which it could be inserted, such as "Evolution and extinction of species" and "Temporary and evolutionary aspects of scientific theories" (Socles de compétences 2001). The official instructions do not therefore advocate teaching natural selection to pupils completing elementary school. Moreover, some teachers feel that it is not possible to address certain evolutionary concepts because they involve the development of too high a level of abstraction (Carette et al. 2009).

11.1.1. *Natural selection*

> Can we doubt [...] that individuals having any advantage, however slight, over others, would have the best chance of surviving and of procreating their kind? [...] This preservation of favourable variations and the rejection of injurious variations, I call Natural Selection (Darwin 1866).

In order to understand the complexity and originality of this proposal, it is necessary to specify the long argument that Charles Darwin (1866) develops in *On the Origin of Species*. For this study, we will refer to the analysis of the concept of natural selection carried out by Ernst Mayr (1989). This does not include the contributions of the synthetic theory of evolution (e.g. the genetic basis of the process), whose level of abstraction may be greater. We have chosen to work on this version of the concept of natural selection because it is more accessible to students at the end of elementary school. However, it presents a less important explanatory field than the one developed in the synthetic theory. This choice will certainly limit the use of our framework to work on some problems intended for higher level students.

Ernst Mayr's analysis of the logical structure of the Darwinian mechanism of natural selection, in a populationist paradigm, shows us that this concept is the result of three inferences built by linking five facts (Figure 11.1).

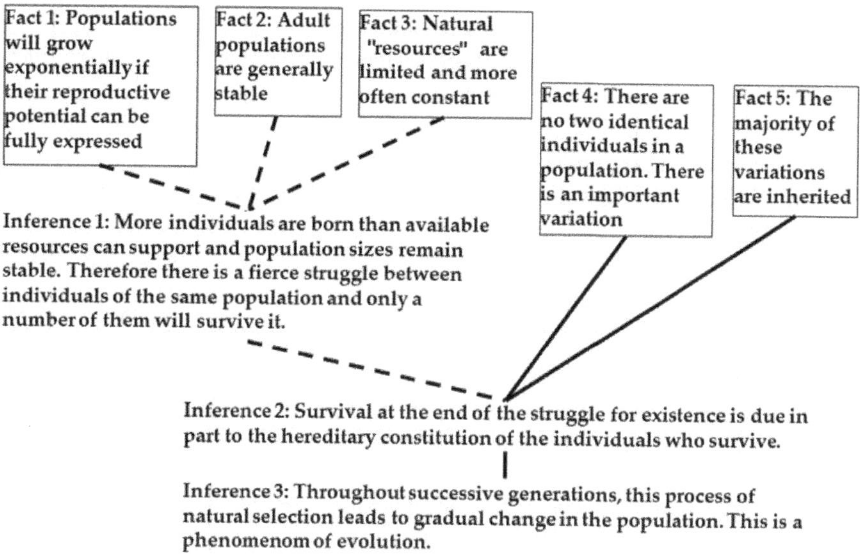

Figure 11.1. *Presentation of the logical structure of the Darwinian concept of natural selection constructed by Ernst Mayr (1989). The dotted lines identify the first causal chain, while the solid lines identify the second causal chain*

This formalization of the concept of natural selection makes it possible to highlight the originality of this proposal in relation to the competing theories of the Darwinian era. The second inference of this concept is constructed from the linkage of two independent causal chains: the struggle for existence (the first inference) and the variability of genetically transmitted characters in the population (the last two facts). This explanation uses several meanings relating to randomness (Gayon 1997), whereas finalist explanations were dominant until then. The construction of this explanation required other reversals of the dominant biological thinking that we will discuss later in this chapter.

The concept of natural selection occupies a very special place in biology as it is at the crossroads of functionalist and historical biology (Mayr 1989). Differential reproduction, restricted to differential survival in Ernst Mayr's formalization presented above, is based on variations in characteristics (e.g. beak width) that participate in biological functions (e.g. nutrition) in a specific context (e.g. an environment offering a large proportion of robust seeds, the consumption of which is easier for those with sufficiently large beaks). The phenomenon of natural selection is therefore intelligible on the condition that we understand some of the functions present in the organism studied and the context in which they take place. In addition, the reconstruction of an evolutionary scenario using the concept of natural selection requires the mobilization of elements from historical biology, such as past events (e.g. changes in the living environment), long time scales and catastrophes (of the second level as established by Denise Orange Ravachol (2005)). If natural selection is not predictive, it does have a retrodictive power because "it allows us to know, retroactively and retrospectively, the material causes that lead to the formation of species" (Gayon 1992).

11.1.2. *Epistemological obstacles...*

To understand the learning process of natural selection, it is useful to refer to the concept of the epistemological obstacle (Bachelard 1938). This can be understood as a double-edged habit of thinking. These habits of thinking are strongly anchored because they are useful and functional for thinking about everyday life and constructing common thought. However, they prove to be problematic when it comes to solving scientific problems. From a historical point of view, the construction of the concept of natural selection required the overcoming of a significant number of epistemological obstacles.

The analysis of some of the barriers to learning about natural selection has already been the subject of several recent publications (Lhoste 2008; Peterfalvi 2008; Fortin 2009; Gobert 2014). This work provides us with the tools to analyze and understand the evolution of students' conceptions during our experiment. We consider the work and the overcoming of obstacles identified among students as one of the main conditions for the construction of the concept of natural selection.

Natural selection is, in the Darwinian sense of the term, an explanatory mechanism for the modification of species over time that applies at least to a population of individuals over several generations. From a historical point of view, the construction of this mechanism therefore implies going beyond fixism, which is based on the immutability of species. However, this overcoming cannot be total as the possibility of the appearance of new variants is constrained by the biology of organisms and, in particular, their reproduction and development. The construction of an explanation of the modification of species by natural selection requires, in fact, going beyond "archaic and primitive thought, which has made and continues to make massive and constant use of metamorphosis, of the conversion of specific animal forms into one another" (Canguilhem 1990). Natural selection must therefore be considered as a mechanism that explains the modification of populations over time under reproductive and environmental constraints.

Fixism, which has been widespread for many centuries, is linked in particular to creationism and typological thinking (Mayr 1989). The latter postulates that each species can be represented by an invariable type to which each of its individuals corresponds. The construction of the concept of natural selection is required going beyond this approach because differential reproduction is exercised on variants present in the population, as Darwin (1866) points out, "these individual differences are highly important for us".

The Darwinian mechanism mobilizes population thinking because it explains the modification of populations over several generations. Each individual carries a series of traits that influence its probability of reproduction in different ways. It is the modification of the distribution of traits over generations that must be explained, and not the reproductive capacity of a typical individual that would be representative of the whole population. In Darwin's case, population thinking was constructed in reference to genealogy, which builds the networks of individuals that develop over the generations.

At the time of the publication of *On the Origin of Species*, some creationists defended and disseminated the idea of harmony between living organisms and their environment (Paley 1802). This representation of reality is an obstacle to the construction of the mechanism of natural selection, whose differential reproduction is based, in particular, on the existence of environmental constraints. It would seem that the reading of *Population* (Malthus 1803) and its qualitative modeling offered Darwin a tool for understanding how "the struggle for existence" could be applied in natural populations. In particular, the realization that there is a struggle for resources between individuals, rather than between groups of individuals, would have played this role (Mayr 1989).

The idea of the existence of a harmony between living organisms and their environment can be, in some cases, an expression of finalism. This is another obstacle to the construction of the Darwinian mechanism. This mechanism can be integrated into a wide variety of explanatory scenarios, such as creationism or Lamarckism. It postulates the existence of a finality in the processes explaining life and its diversity. Darwin (1868) overcame this obstacle, notably by recognizing the independence of the effect of new variations from the constraints of the environment in which they appear: "the variations of each creature are determined by fixed and immutable laws; but they have no relation to the living structure which is slowly built up through the power of selection". His argument mobilizes, in this way, the concept of randomness already evoked in the formalization of the mechanism of natural selection by Ernst Mayr. Natural selection thus corresponds to the meeting of two independent causal chains.

Without adhering to the idea of a homology between the students' thought processes and the historical development of biological concepts, the analysis of these obstacles in the course of the history of biology allows us to build a framework for constructing and analyzing our research.

11.1.3. ...*to students' conceptions*

The interest of our study for research in evolution didactics is based in part on situations proposed to students around 12 years old. Until now, very little research in French has been done on conceptions of natural selection and the dynamics of their modification in young students. The state of research on the issue of teaching natural selection mirrors the reality of the classroom, as few teachers report teaching it in the Wallonia-Brussels Federation (Carette et al. 2009).

To the best of our knowledge, there is no French-speaking Belgian study that looks at primary school pupils' conceptions of the mechanisms of evolutionary theory. In France, we will focus on the work of Patricia Crépin (2002) and Corinne Jégou (2009). First of all, it appears that no primary school student is able to cite or mobilize the concept of natural selection at this level of education. Several explanatory systems are identified in their responses, but no student seems to mobilize any of these explanatory systems. There is, however, a certain coherence in the students' conceptions as they preferentially associate fixist, spontaneist[1] and creationist systems, on the one hand, and transformist and dinosaurist systems[2], on the other.

The conceptions mobilized by the students contain explanations that can be directly related to the obstacles cited in the epistemological analyses concerning natural selection. Table 11.1 lists the conceptions identified among elementary and secondary students and the barriers they may mobilize. Some of these obstacles may also have been documented at different grade levels.

Among students who exhibit transformist thinking in the broad sense[3], there is a tendency to think at the individual level rather than at the population level. Patricia

1. System mobilizing an explanation close to the theory of spontaneous generation.

2. System indicating that dinosaurs were the first organisms to appear on earth and other animals are their descendants.

3. We choose to use the term "broad transformism" to broaden the range of student explanations that can be included in this category. Transformism is frequently associated with Lamarckism, whereas, as Patricia Crépin (2001) acknowledges, students most often produce transformist explanations that depart from the Lamarckian model. As an example, this

Crépin (2001) explains this difficulty as a "lack of structuring of the notion of species". The transformations identified by the students seem most often to take place during the life of the individuals. Another confusion appears between the notion of the life cycle and the notion of successive generations. The pupils would present a lack of structuring of the notions of ancestors, descendants and family trees.

L	Conceptions	Possible obstacles
E	Spontaneous: species can appear without parents according to the idea of spontaneous generation (Crépin 2002)	- Vitalism (Fortin 2009)
E	Dinosaur: Dinosaurs were the first to appear on Earth and current species may be their descendants (Crépin 2002)	- Fixism/essentialism
E	Fixist (Crépin 2002)	- Fixism/essentialism (Crépin 2002; Coley 2015) - Cyclic/anhistoric time
E and H	Creationist (Fortin 1993) (Crépin 2002): species were created by God Concordist (Fortin 2009): mixture of creationism and scientific concepts	- Creationism (Crépin 2002; Uyen 2013) - Fixism/essentialism - Finalism (Crépin 2002; Uyen 2013) - Cyclic/anhistoric time
E and H	Transformism in the broad sense: species or individuals can transform themselves with little or no constraints (Fortin 1993) (Crépin 2002) (Carette et al. 2009)	- Metamorphosis (Brumby 1984; Fortin 1993; Crépin 2002; Uyen 2013) - Vitalism (Bishop and Anderson 1990; Fortin 1993; Uyen 2013)
H	Non-evolutionary: independent lineages exist and can improve (Fortin 2009)	- Fixism/essentialism (Fortin 2009) - Finalism (Fortin 2009)
H	Adaptationist: only the action of the environment explains the evolution of species (Coquidé 2009)	- Finalism - Environmentalism: the environment is seen as a mutagen (Bishop and Anderson 1990; Fortin 2009)
E and H	The obstacle is not attached to any particular design.	- Competitive: evolution is driven by the law of the strongest (Fortin 2009).
E and H	The obstacle is not attached to any particular design.	- Kinship barrier: evolutionary kinship made of genealogical cleavage is confused with family kinship made of genealogical continuity (Bishop and Anderson 1990; Fortin 2009)

Table 11.1. *Key conceptions and barriers identified among students in research on teaching-learning evolution. E = elementary, H = high school. Essentialism is the school of thought that underlies typological thinking*

category includes the pseudo-evolutionist and transmutationist conceptions identified by Corinne Fortin (2009).

Some of the students interviewed consider that environmental constraints are the only causes explaining the transformation of individuals over time (Crépin 2001). They therefore do not consider the idea of intraspecific variation as the source of transformations that can then be selected by the environment. Other students develop a form of Lamarckian finalism in which organisms transform themselves in response to environmental constraints (Crépin 2001). Finally, a form of vitalism can also be observed, in which students analyze "the adaptation of organisms as an immediate response to their needs" (Fortin 2009).

Students' conceptions of the evolution of species and the mechanisms that explain it are very diverse. A majority of them mobilize obstacles similar to the epistemological obstacles identified in our epistemological analysis. Our work questions the capacity of students at this level of education to develop conceptions closer to the Darwinian mechanism of natural selection. This research therefore involves understanding and determining the conditions that could allow the development of these new conceptions.

11.1.4. *The value of problematization in understanding the learning of natural selection*

The modification of the conceptions of 6th-grade students concerning the processes that explain the evolution of species must involve the obstacles already presented in the preceding analysis. In order to carry out this work, we agree with Michel Fabre and Christian Orange (1997) that "it is above all the construction and reconstruction of problems by the learner that must be studied by the didacticians and worked on in class" if we wish to establish "the conditions that will allow a real overcoming of obstacles as well as the adhesion to another conception".

The problematization paradigm (Orange 2012; Fabre 2016) developed by its two authors offers a theoretical framework and analytical tools for constructing a didactic intervention and studying the processes of problem construction and explanations that take place. This paradigm, based on Gaston Bachelard's (1938) epistemological approach, models scientific activity as a work of explanatory problems.

In this model (Figure 11.2), scientific activity is defined as the construction of an interaction between a register of facts and experiments, called the empirical register, and a register of explanatory ideas, called the model register, by a scientific community sharing a paradigm and possibly other theoretical elements, called the explanatory register (Orange 2012). This modeling is represented as a constraint space. Without assuming the existence of a homology between scientific work and classroom work, this model can be used to analyze a classroom intervention aimed at having students construct a problem and explanations that could answer it.

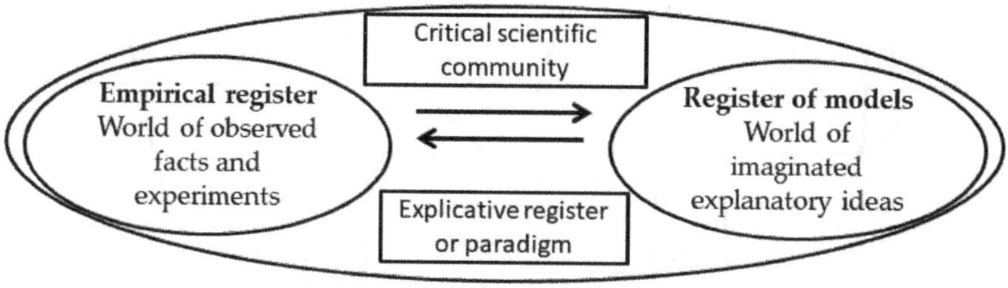

Figure 11.2. *Diagram of scientific activity as explanatory problem work according to the problematization paradigm (Orange 2012)*

This model reinvests the assertoric / apodictic couple[4] by recognizing that scientific explanations belong to the second category. They are different from the propositions of common thought because they are reasoned and must be articulated to the theoretical framework in which they are produced. They are then identified as necessities and they are constructed from a "critical relation of elements of the empirical register (of facts) recognized as relevant and elements of the explanatory register" (Orange 2012).

The theoretical framework of problematization was mobilized to study how students take on a problem concerning the evolution of a population. The Darwinian mechanism of natural selection was thus represented in the form of a space of constraints (Figure 11.3). This was constructed from the analysis of the logical structure of the Darwinian concept of natural selection carried out by Ernst Mayr and presented in the introduction. This was used as a template for constructing our didactic intervention and for analyzing the students' productions.

4. A knowledge is assertoric if it is not related to a theoretical system (e.g. when dissecting a digestive system, we can see that food is transformed there). On the other hand, knowledge is apodictic because it maintains a relationship of necessity with a given theoretical framework (e.g. in a mechanistic theoretical framework, food must be reduced as it passes through the digestive tract to make it possible for it to pass into the blood) (Orange 2012).

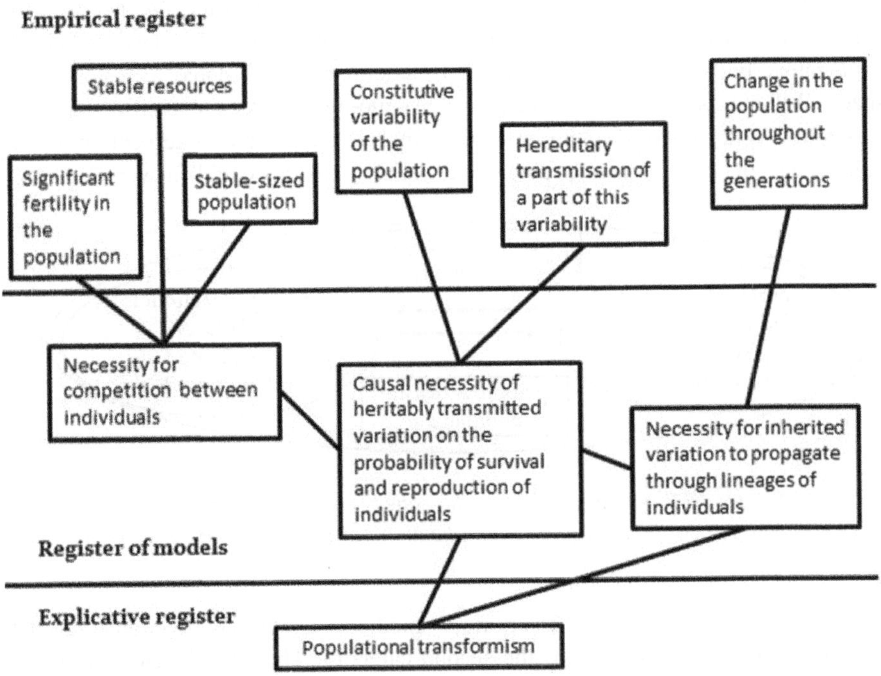

Figure 11.3. *Reference space of constraints of the Darwinian problematization*

11.2. Analysis device and methodology

Our research is a case study within the paradigm of feasibility research because it is aimed at identifying some of the possible problems and explanations that are constructible by 6th-grade primary students that are "difficult to imagine a priori" (Astolfi 2003). The research was carried out through the partial mobilization of the methodology of enforced situations (Orange 2010). Enforced situations are teaching situations co-constructed by the teacher and the researcher that aim and control, respectively, pedagogical objectives (teaching) and research objectives (producing phenomena). The teacher must have been involved in the research for some time and participates in the a priori analysis, as well as in the choice of situations. This methodology recognizes the partially contingent nature of classroom functioning and takes it into account by allowing for adaptation during the sequence. In this framework, the comparison of the a priori analysis and the a posteriori analysis allows for the creation, identification and analysis of new didactic events (Orange 2010).

This research involved constructing a didactic intervention, administering it to a class of 6th-grade students and analyzing the productions collected during the intervention. Firstly, we will detail the content of this intervention and then we will develop the analysis methodology used. Finally, a part of the analysis of the oral and written productions will be presented and discussed.

Our didactic intervention took place over the course of seven sessions, shown in Figure 11.4, which varied in length from 30 minutes to 2 hours. Each session was videotaped, and interactions between the teacher and students were audio recorded. The purpose of this recording device was to capture all of the teacher's and the students' words exchanged in large groups.

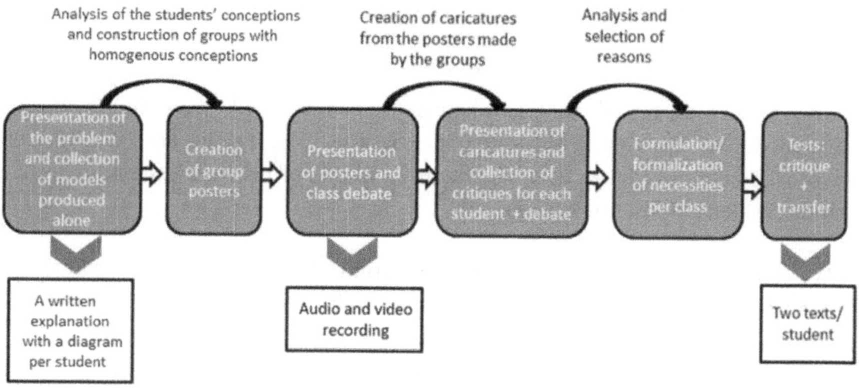

Figure 11.4. *Composition of the didactic sequence experienced by the students and identification data used to carry out our analysis. The gray boxes identify the sessions and their content. The white boxes identify the data collected and presented in this chapter. The curved arrows show the actions taken by the teacher and the researcher between certain sessions*

In session 1, students were presented with a problem concerning the evolution of several populations of peppered moths (see below) and were asked to write an explanation in response.

During the second session, students from several groups had to produce a poster with a common explanation.

The third session was dedicated to the presentation of the posters and the debate of the proposed explanations. Following the analysis of the explanations already developed, the teacher and the researcher produced three caricatures[5] that synthesize the majority of the students' proposals.

In session 4, students were asked to critique these caricatures, in writing and individually, before sharing and debating with the class.

The fifth session was dedicated to the categorization of certain sentences produced individually by the students in order to collectively reconstruct a text of knowledge. Two new caricatures were offered to the students in the following session. They must again produce a critique individually and in writing.

Finally, the last session was dedicated to the presentation of a problem similar to the initial one and to which the students had to propose an individual explanatory answer.

The results presented in this chapter are constructed primarily from data collected during the first, third and final sessions. The results are derived from individual student responses at the beginning and end of the intervention, as well as from discussions in front of the class.

We borrow our analysis methodology from the theoretical frameworks of problematization (Orange 2010). An a priori analysis of our didactic intervention was carried out in order to control the meaning that will be attributed to the students' and the teacher's behaviors during the a posteriori analysis. This a posteriori analysis of our didactic intervention was carried out on the basis of the transcripts of the oral productions shared by the students and the teacher and on the basis of the students' written productions. They were analyzed using tools from the theoretical framework of problematization which allow us to establish the argumentative activity and the epistemological components. The latter were identified by using, as a reference, our epistemological analysis of natural selection presented at the beginning of this chapter. The analysis of the argumentative activity is carried out thanks to Plantin's trilogal model (1996), Toulmin's monologal model (1993), the

---

5. To get students to produce written arguments, it is possible to use caricatures. These are diagrams that are offered to students and that "take the basic functional characteristics of some group productions by homogenizing them in terms of form and depersonalizing them" (Orange 2012). In the case of our study, it was necessary to greatly evolve the initial form of the students' schemas because few of them represented the orally evoked processes with sufficient accuracy.

negotiation of schematizations (Orange 2003) and the construction of constraint spaces (Orange 2000)[6].

11.2.1. *Didactic intervention*

Our didactic intervention was built from the well-documented example of the peppered moth, *Biston betularia*, developed by Kettlewell (1959). It was decided that an example of reduced complexity would be chosen in order to favor the problematization and eventual construction of the Darwinian concept of natural selection. While this case study is subject to some criticism (Majerus 1998), it is nonetheless an excellent case study for classes (Majerus 2008). It has the advantage of being characterized by a relatively simple set of parameters for a natural selection case, while being located in an environment potentially familiar to students:

– the phenotypic character has two forms, white or black (morpha typica and morpha carbonaria respectively), determined by a monogenic genetic system with two alleles;

– environmental variation is explained by a simple causal chain that seems understandable for 6th-grade elementary students.

The didactic intervention begins with the presentation of a problem to the students, the description of which is shown in Figure 11.5. This problem was constructed by the researchers and the teacher.

Students were asked to explain the change in a population of peppered moths in two woods between 1830 and 1880 knowing that:

– In 1845, factories were set up next to one of the two woods studied. These factories emitted smokes that blackened the trunks of the birch trees on which the moths lived.

– The moths were counted from 1830 to 1845, during which time only white moths were counted, and in 1880, when 70 black moths and 30 white moths were found in the woods with factories, as well as 99 white moths and one moth of another color in the woods without factories. During the oral presentation, it was specified that a black moth appeared in the wood without a factory in 1835 and that no black moth was found in this same wood in 1836.

– Moths are hunted by "sight-hunting" birds.

6. The reader can find an explanation of the methodology in the book by Christian Orange (2012).

To these elements, information was added which made it possible to specify several dimensions of the problem. The environmental constraints were first clarified by presenting an image of two moths of different colors on two different types of trunks. If students associated this information with that of the predators, they could construct the principle of differential predation. The second environmental constraint concerned the distance between the two woods, making it impossible for moths to immigrate or emigrate.

The other information presented in the didactic intervention concerns the modalities of reproduction of moths. The hereditary transmission of colors and the relative quantities of the offspring of each color in each type of pair were specified to avoid the mobilization of the model of heredity by mixing (Rumelhard 1986), which would risk overcomplicating the problematization process. It was stated orally that each female would produce 20 viable eggs that would hatch in the spring of the following year. This clarification allowed students to structure their conception of the moth's life cycle while becoming aware of the population's capacity to grow thanks to its reproductive potential.

It was finally indicated that a moth of another color could appear every thousand moths. This possibility was put into practice through the appearance of a black moth in the wood without a factory in 1835. This information was offered to the students so that they could mobilize the principle of inter-individual variation. The application of this possibility to the wood without a factory could be used to put to work the vitalist obstacle[7] identified in our epistemological analysis. The set of elements concerning the modalities of the reproduction of moths could finally allow the students to enter into population thinking.

The construction of the problem by the students could also be based on the tasks proposed to them. In the first session, they were asked to produce individually an explanatory text and a schematic representation of the proportions of moths of different colors in the populations of 1830, 1865 and 1880 (see Figure 11.6). The point of this request was to encourage their entry into population thinking and to obtain a complementary development of their explanation of the process of population change over time.

7. Students could use the case of this moth to challenge a proposal that white moths transform into black moths to survive against sight-hunting birds.

Problem summary sheet

The peppered moth:

- lives on the trunks of birch trees,
- is eaten by 'sight-hunting' birds when they are resting on the trunks,
- reproduces once a year and in the following ways:

A) 1 white moth + 1 white moth => white moths
B) 1 white moth + 1 black moth => some white moths and some black moths
C) 1 white moth + 1 black moth => few white moths and lots of black moths.

Warning! One moth in 1,000 is born a different color to their parents.

One moth produces lots of offspring

The problem:

From 1830 to 1845, there were more or less always 100 white moths in each of the woods studied.

In 1845, factories were built next to one of the woods. Their smoke blackened the birch trunks that were found there.

In 1880, only 30 white moths per 70 black moths were found in the woods next to the factories. In the wood without factories nearby, 99 white moths and 1 different colored moth were found.

First picture : Black peppered moth and white peppered moth on a birch trunk before 1845.

Second picture : Black peppered moth and white peppered moth on a smoked birch trunk after 1880.

Figure 11.5. *First part of the summary document (translated into English) of the didactic situation distributed to the students. This is a document produced by the researcher and the teacher (Kettlewell, 1956)*

1. How can you explain that the group of peppered moths from the wood with factories nearby was composed of lots of black individuals and very few white ones in 1880, whilst in 1830 it was composed only of white peppered moth?

2. DRAW, on the back of the sheet, one or more figure(s) which explain your response to the question.

To do this, represent schematically the ratios of different colored moths in:

-the 1830 population;
-the 1865 population;
-the 1880 population and;
-the events which allow you to explain the population changes over time.

To represent the groups of these three populations, represent the peppered moth as indicated below. To help you, templates or pre-cut the peppered moths are available.

A white peppered moth:  A black peppered moth: 

Figure 11.6. *Written presentation of the problem and the instruction distributed to students. In the original version, there was a 13-line space for students to propose an initial answer between the question (1) and the diagram instruction (2). This part was located at the beginning of the second page distributed to the students*

In the second session, student groups were asked to produce an explanatory text and a poster to illustrate their explanations. The creation of this poster was constrained by certain instructions presented in Figure 11.7, which encouraged the students to model the two moth populations of 1865 in the form of five reproductive pairs for each of them. They were then asked to represent the offspring produced by these five pairs and the adults they became in 1866. The objective of this modeling was to allow the students to mobilize the reproductive potential of the population and to eventually develop the problem of the necessary annual reduction of the moth population. Indeed, the annual reproduction of the population made it possible to produce 1,000 young moths, whereas only 100 adult moths were counted each year. This modeling could therefore also promote the entry into population thinking.

For the year 1865...

- transform the 1865 population of 100 moths into a population of 10 moths
- establish 5 couples using the population of 10 moths
- represent for each of these couples, the 'offspring' that will be produced
- represent the adults (ready to reproduce) that will have these offspring in the 1866 population
- represent the 1880 population

Write a text describing the explanatory model that you have produced to respond to the first question

Figure 11.7. *Presentation of the additional instructions for group poster construction. This instruction was distributed to each group at the beginning of the second session*

	Initial problem: moths	Final problem: the mice
Starting situation	Exclusive presence of white moths (W) living on white birch trees (WS)	Exclusive presence of sand-colored mice (W) living on sand (WS)
Selective factor	Chickadees hunting on sight (P)	Raptors hunting on sight (P)
Environmental change	The trunks of birch trees turn black (BS) due to the release of smoke (CA) from the factories	Part of the sand is covered by hardened black lava (BS and CA)
Time scale	45 years old	300 years
Reproduction	10 individuals per laying with qualitative details on the results of the different reproductive combinations	10 individuals per laying **without** qualitative details on the results of the different reproductive combinations
Distribution	Two initially identical populations in two very distant woods. Only one population undergoes environmental modification and natural selection.	There is initially only one population. It is subject to environmental modification in part of its territory and the phenomenon of natural selection.
Problem	How is it possible to explain the change in population in the woods next to which the plants are located?	How is it possible to explain the appearance of many black mice on the new black earth crust?

Table 11.2. *Comparison table of the two problems proposed to the students*

Finally, the students were asked to propose an explanation to a new problem at the end of the didactic intervention. This new problem was constructed to be similar to the previous one. It is not possible to present it in detail in this chapter. We can nevertheless mention the similarities and differences between the two problems (Table 11.2). For ease of presentation of the results, we will use the following terms to describe the two situations: a white individual (=W), a black individual (=B), the predators (=P), the coloring agent (=CA), the white substrate (=WS) and the black substrate (=BS).

11.3. Results

11.3.1. *Individual student explanations at the beginning and end of the sequence*

Analysis of the students' individual responses to the problem presented at the beginning of the didactic intervention (see the first column for each student in Table 11.3) indicates the presence of five groups whose explanations differ with respect to the presence of one or more black moths. It was decided to categorize the students' responses based on this component of their explanations because it is essential for the construction of the Darwinian concept of natural selection. These five groups are not directly visible in the table because it was organized according to the answers provided to the problem posed at the end of the didactic intervention:

– The most common explanation offered (8 out of 19 students) indicates the darkening of white individuals by the coloring agent (smoke).

– The second largest group of responses (4 students) aggregates explanations mentioning either the adaptation of the individuals or the will of the white individuals to become black following the blackening of the white substrate by the coloring agent. The majority of these students mention the increased visibility of white individuals in this new environment and some of their proposals reflect the implementation of the metamorphosis obstacle[8].

– The third group of responses (3 students) combines the explanation of coloration by smoke with the explanation of the moths' desire to become black to camouflage themselves. As an example, student #7 offers the following explanation:

---

8. Here, as an example, is a student proposal (spelling retained) that reflects the implementation of this obstacle: "The moths naturally took on the color of the birch trees but how? You will tell me, they simply adapted to the birch tree" (ST 3). This proposition reflects the ability of white individuals to "naturally" take on a new color.

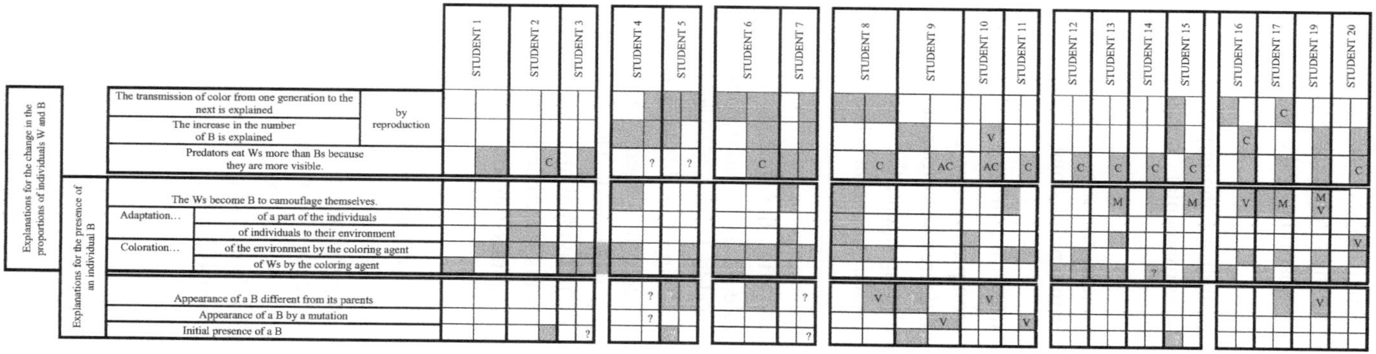

Table 11.3. *Comparative table of components of written explanations produced individually by students in response to the problem posed at the beginning of the instructional intervention and in response to the problem posed after the instructional intervention was completed. B = a black individual, W = a white individual. For the explanations provided in response to the problem proposed at the end of the didactic intervention, a code with letters was applied: a doubt on the determination of the component is marked by a "?", "V" indicates the presence of the vitalist obstacle, "C" indicates the development of a caricatural vision of the phenomenon, "M" indicates the presence of the obstacle of metamorphosis and "A" indicates the presence of the anthropomorphic obstacle*

The black smoke of the factories colored the food of the moths and when the moths ate their food they became black. As the birch trees became black, the moths, in order to hide, had to become black by adapting to their environment with their cells[9];

– another group gathers students who indicate the appearance of a black moth without specifying the manner or environmental context that allows this phenomenon to occur;

– finally, the last group mentions the initial presence of black moths when this is not consistent with the counts proposed in the initial situation.

The explanations proposed in each of the five groups may vary in their level of complexity. Some of them mobilized complementary elements to those already mentioned. We find among these and in a decreasing order of representativeness: the coloration of the environment by the coloring agent, the rules of reproduction, the death of white individuals and the differential predation which are exerted on white individuals and on black individuals.

The profile of the students' answers after the didactic intervention (see the second column for each student in Table 11.3) is more complex since we have identified six groups of explanations in Table 11.3. A first group of three students explains this new phenomenon by mobilizing the differential predation and the blackening of the environment by the coloring agent without considering the modalities of appearance of a first black individual. The absence of explanation concerning the appearance of a first black individual also characterizes the answers proposed by the second group. These answers have been gathered together because they also mobilize the principle of transmission of the color black through reproduction.

The third group is composed of two students who presented conceptions bringing together some elements that constitute the Darwinian concept of natural selection. Among the common points, we find the coloration of the environment, the heredity of the color of individuals and the presentation of the principle of differential predation, even if this is a caricature of one of the two students, as will be shown by the comments of student #6. The two students indicated the rare appearance of a black individual but one of these students did not mention the origin of this individual. The words of student #6 help to illustrate the characteristics of the conceptions of this third group:

9. In this case, the differential predation is only visible on the diagram made by the student.

> At the beginning, there are only white individuals […]. A mother has produced 1 black individual […] (this happens very rarely). When the substrate […] has become black, the white individuals […] cannot hide well. So they are eaten by their predator more easily. On the other hand, the black individual […] can hide very well because the substrate […] became black. This black individual […] reproduced and made black individuals […] and its black individuals […] reproduced again […] the blacks were not eaten. On the other hand, the white individuals […] were eaten because they did not know this camouflage. So there was a decrease in white individuals […] and an increase in black individuals...[10]

The fourth group is characterized by the proposal of a principle of differential predation, but often expressed in a caricatural or anthropomorphic way; the individuals choosing, in particular, to camouflage themselves on the grounds of which color makes them less visible. Moreover, the birth of a black individual from a pair of white parents or through mutation is evoked. A form of vitalism is however detected in the chosen expressions. The mention of mutation was incorporated in the preceding sessions as a result of suggestions from students and the teacher. The explanation of what a mutation is, however, was never developed. Student #9's explanation provides an account of the characteristics of this group's responses:

> Black individuals […] camouflage themselves on the black substrate... The white individuals […] camouflage themselves on the white substrate... The predators eat the black individuals […] on the white substrate […] and the predators eat the white individuals […] on the black substrate... The black individuals […] came through a mutation is a strange reproduction.[11]

The penultimate group is identified by sharing the following explanations: differential predation is recognized sometimes in a caricatural way, the survival of black individuals being total. In parallel, the appearance of black individuals is explained by the same factor that colors their environment. Three students out of four nevertheless add the will of the individuals to camouflage themselves. The last group also presents in their explanations the principle of differential predation and the will of the individuals to camouflage themselves. Reproduction is often recognized as an element that increases the number of black individuals. Two

---

10. The problem-specific terms have been replaced by generic terms that may be appropriate for both problems. The latter are each time identified by the addition of the following mark: "[...]".

11. Ibid.

students out of four finally take up the possibility of the appearance of a black individual from two white parents.

More generally, we can thus note the almost permanent mobilization of a principle of differential predation. However, this principle is partially mobilized in a caricatured way, indicating most often the predation of white individuals only. The appearance of a black individual following a reproduction event is proposed by almost half of the students. This phenomenon is however sometimes associated with vitalism, either directly in this explanation or in the explanations that accompany it. The desire of individuals to camouflage themselves appears in the words of seven students, marking in this way the importance of vitalism in the students' explanations. The ability to transmit the color black and, to a lesser extent, the increase in the number of black individuals through reproduction were mentioned by more than half the students. These explanations are present in groups of students who systematically look to the appearance of a black individual through reproduction.

A comparison of the explanations provided by the students in response to the first and second problems reveals a significant change in their conceptions. Analysis of the response elements mobilized by the students allows us to identify the following trends:

– the decrease in the number of students (from 11 to 6) using the explanation of the appearance of black individuals thanks to a coloring agent from the environment (smoke or dust depending on the problem);

– the stagnation in the number of students (5) indicating that "white individuals become black to camouflage themselves". The identity of students making this choice, however, varies between the first and second problems. This stagnation does not, therefore, reflect a stagnation in students' conceptions;

– the increase in the number of students (from 7 to 13) mobilizing "the change in color of the substrate on which the populations live";

– the increase in the number of students (from 1 to 17) indicating, often in a caricatured manner (11), that "predators preferentially eat white individuals over black individuals because they are visible";

– the increase of pupils (from 0 or 2 to 8) mobilizing "the appearance of black individuals following the reproduction of two white parents and/or thanks to a mutation". As the initial answers of the students were not always explicit in relation to this point, it is difficult to know if this element had already been mobilized in the face of the first presentation of the problem.

11.3.2. *Dynamics of the construction of the concept of natural selection*

The interest of our study is to document and deepen the understanding of the dynamics of the construction of the Darwinian concept of natural selection by a class of 6th-grade elementary students. To do so, it is necessary to analyze the debates that took place between the initial presentation of the problem (first session) and the construction of a knowledge-based text focused on the Darwinian concept of natural selection (fifth session). The overall dynamic of the construction of the Darwinian concept of natural selection is long and complex because it consists of many back-and-forth discussions on the different themes identified during the main debate. We present only some parts of this dynamic, all referring to the peppered moth problem since they take place during the working phase of this problem.

It can be seen that the idea that black individuals appear thanks to a coloring agent coming from the environment is less mobilized by the students in the second problem. The questioning of this coloring mechanism takes place at the beginning of the debate. The highly argumentative moment presented in Figure 11.8 allows us to see two objections proposed by the students.

The first objection is based on the idea that smoke staining cannot explain the appearance of a black moth in wood without a factory (ST 2, intervention 108). This idea will not be repeated later. This observation questions the students' ability to include the second wood case in the problem. The second objection (ST 5, intervention 117), on the other hand, is much more effective since it is repeated six times by several students at different points in the debate. It is built on the idea that smoke coloring cannot work because smoke is not a natural color. However, according to the students, only a natural color can be passed on to descendants. At the time of the discussion, this objection seemed to be unanimously accepted by the class. We therefore consider that it can take on the status of a presupposition, i.e. that it is an element of the explanatory register of the class that has already been stabilized before the activity.

From the point of view of problematization, this presupposition may have a second essential function. It makes it possible to mobilize "the need for heritable variation to propagate in lineages of individuals" that composes our model of the problematization process of the Darwinian concept of natural selection (Figure 11.1).

During this first part of the debate, the class was put to work and stabilized the annual production model of 1,000 moths presented in our analysis a priori. The instructions for constructing the posters and the joint work of the teacher and the students thus enabled the latter to mobilize the principle of the population's reproductive potential. The intention of having the students develop this model is explained by the desire to engage them in solving the problem of the annual reduction of the moth population and, eventually, the differential reduction of moths in the population. The discussion reflects the realization of this intention as the fourth topic, which addresses this problem, is strongly represented there.

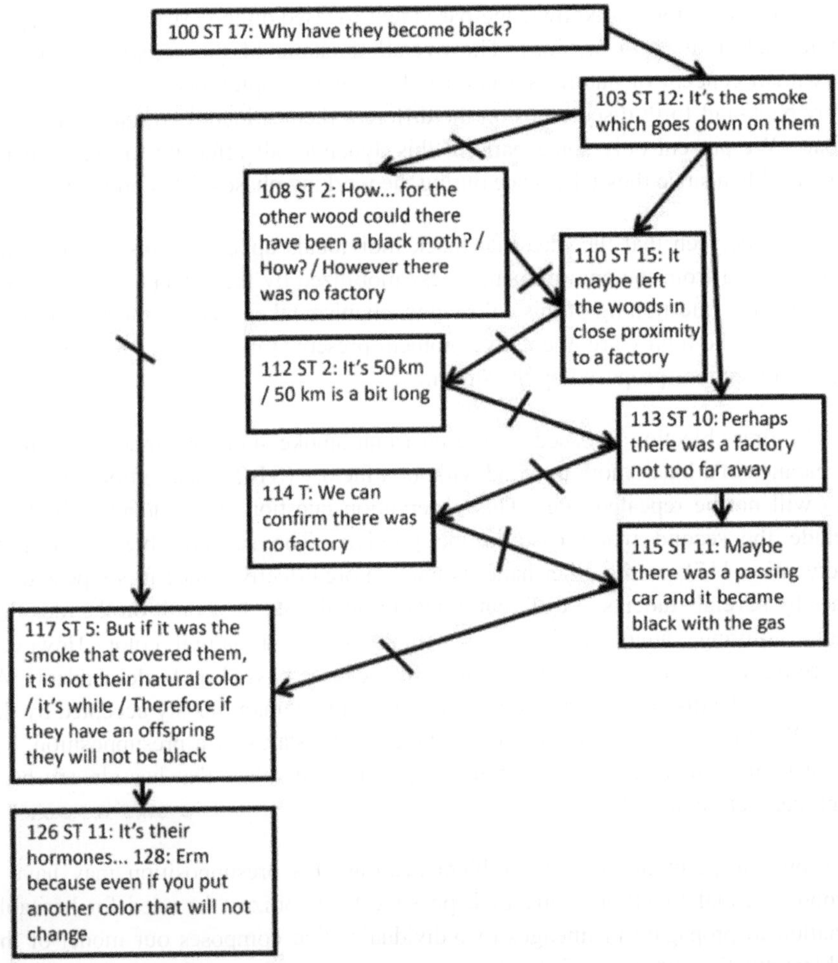

Figure 11.8. *Argumentative structure of the highly argumentative moment dealing with the possibility of smoke staining butterflies. ST = student, T = teacher. The number at the beginning of each intervention corresponds to the chronological number of the verbatim in the debate. Simple arrows mark the continuation of an idea or the reaction between two interventions. Crossed-out arrows mark an opposition between the two interventions*

The processing of this problem by the class and the mobilization of the idea of the camouflage of moths of the same color as their substrate (Table 11.4) allowed the students to partially construct the "causal necessity of the variation transmitted in a hereditary way on the probability of survival and reproduction of the individuals" (Table 11.5). The idea of differential predation was acquired by a large part of the class since 17 students mobilized it, sometimes in a caricatural way, in their answers to the second problem (see the second column for each student in Table 11.3). From the point of view of problematization, the idea of differential predation completely replaces "the need for competition between individuals". The massive mobilization of this principle could also be spotted in an American study using a device close to the one present in our research (Metz 2010).

93	T	Why did they have to change color? What if you insert a black background? With dark birch trees. What would that show at that time?
94	ST8	Well, if they changed their color to black to camouflage themselves.

Table 11.4. *Excerpt from theme 6 illustrating the mobilization of the idea of camouflage. ST= student, T = teacher. The number at the beginning of each intervention corresponds to the chronological number of the verbatim in the debate*

465	T	... And we try to build together to find a solution and say where are the 1,000 when every year we count 100?...
...	...	...
468	ST15	Well, there may be many that are seen.
469	ST2	Especially since many of them were not black like the trees and so the predators came and ate them.
470	T	Have they eaten them?
471	ST2	900

Table 11.5. *Excerpt from theme 33 illustrating the application of the principle of differential survival in the debate*

In most cases, the mobilization of the camouflage principle and, to a lesser extent, differential predation in a caricatural way is associated with the idea that the first black moth appeared voluntarily in response to new environmental constraints (Table 11.5). These propositions demonstrate the engagement of the vitalist obstacle and the metamorphosis obstacle. Intervention 94 of student #8 in Table 11.4 suggests that all the moths changed color to camouflage themselves. They thus appeared to be an expression of strict determinism that prevented the development of a probabilistic, population-based view of natural selection (Gobert 2014). Nevertheless, it was possible to identify in some students (6 students) traces of a first probabilistic approach to natural selection: "For white individuals [...] it's not easy to camouflage so there is more risk that white individuals [...] will get eaten" (ST18).

Following the objections made at the beginning of the debate against the possibility of the coloration of moths by smoke, an important part of the rest of the debate (themes 7, 8, 9 and 12) concentrated on the search for an explanation for the appearance of black moths. Three themes (7, 8 and 9) were dedicated to the working out of explanations proposed by the students, which were each time expressed by a single word: transformation, evolution and adaptation. The teacher intensively questioned the meaning of these three words. This work allowed the debate to continue and new explanations to be sought, and it therefore seemed essential to the collective problematization process.

It was not until halfway through the debate that a student referred to the appearance of the first black moth following the reproduction of two white moths. The use of this rule, initially present in the summary sheet of the activity, occurred only three times during the debates. It was the search for an explanation for the presence of a black moth in the wood without a factory in 1835 that allowed this mobilization. The case of this wood, which seemed to be out of the question during the processing of the previous necessities, turned out to be an interesting tool to put to work the most difficult element to mobilize.

In the case of a wood without a factory, the appearance of a black moth had no particular function any longer. The mobilization of this explanation, present in the initial elements of the problem, was however opposed by several objections (Figure 11.9, ST13, intervention 778). It thus seemed very difficult for these students to mobilize "the appearance of a black individual following the reproduction of two white moths" in order to finalize the construction of "the causal necessity of hereditary variation on the probability of survival and reproduction of individuals", which is part of the Darwinian concept of natural selection (see Figure 11.1). Among the eight students mobilizing the appearance of a black individual in the reproduction of two white moths, the weight of some of these obstacles was still very much present (Table 11.3). The Clarke-Midura et al. (2018) study also notes the profound difficulty of getting students to move beyond this type of reasoning.

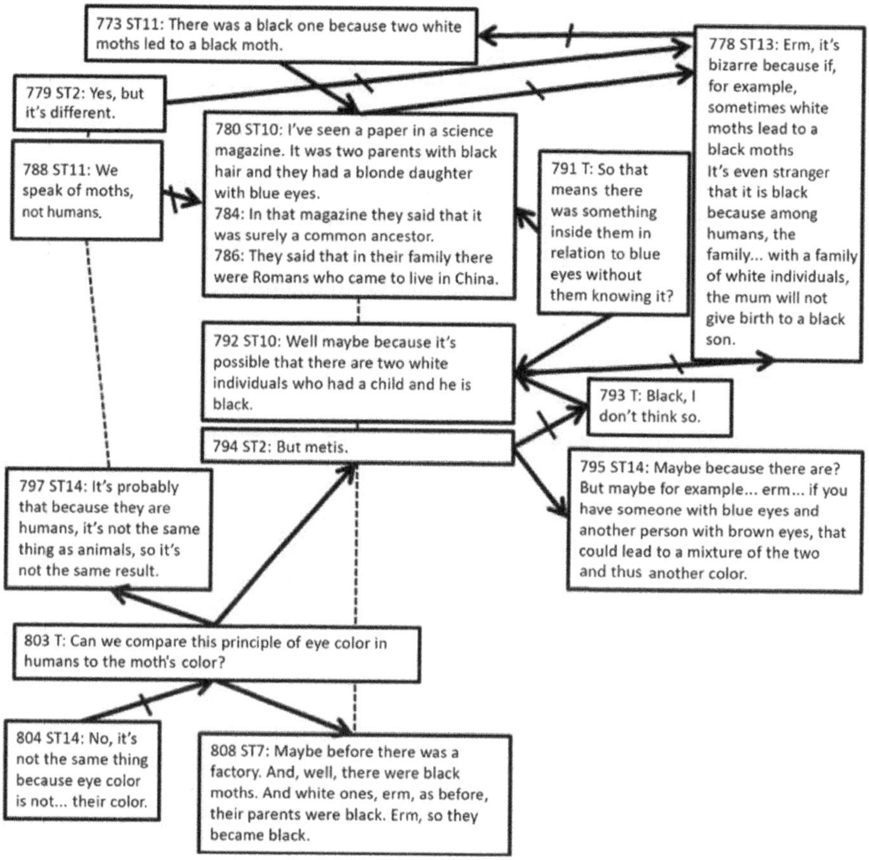

Figure 11.9. *Argumentative structure of the highly argumentative moment dealing with the appearance of a black moth following the reproduction of two white moths. ST = student, T = teacher. The number at the beginning of each intervention corresponds to the chronological number of the verbatim in the debate. The simple arrows and dotted lines mark the extension of an idea or the reaction between two interventions. Crossed-out arrows mark an opposition between the two interventions*

11.4. Conclusion

This research has made it possible to develop and analyze the dynamics of problematization and the construction of explanations concerning the Darwinian concept of natural selection. From the proposed intervention, it appears that the mobilization of the transmission of hereditary traits and differential predation are quite possible for 11–12-year-old students. On the other hand, it is difficult for them to mobilize the appearance of an individual variation independently of its usefulness within a given selective context. These results call for further research to validate these findings in other contexts (Lhoste and Gobert 2022 accepted). It also invites further exploration of students' ability to think about the construction of explanatory scenarios from two independent causal chains. To do so, it is necessary to question students' ability to conceive the appearance of an individual of a new form as an independent event. This approach could be based on the production of several explanatory scenarios, making it possible to vary the moment of appearance of the useful variant. It could also mobilize scenarios in which not all populations persist.

11.5. References

Astolfi, J.-P. (1993). Trois paradigmes pour les recherches en didactique. *Revue française de pédagogie*, 103, 5–18.

Bishop, B.A. and Anderson, C.W. (1990). Student conceptions of natural selection and its role in evolution. *Journal of Research in Science Teaching*, 27(5), 415–427.

Brumy, M.N. (1984). Misconceptions about the concept of natural selection by medical biology students. *Sciences Education*, 8(4), 493–503.

Canguilhem, G. (1990). VIE. Encyclopœdia Universalis. *Encyclopœdia Universalis*, 23, 546–552.

Clarke-Midura, J., Pope, D., Maruca, S., Abraham, J.K., Meir, E. (2018). Iterative design of a simulation-based for teaching evolution by natural selection. *Evolution: Education & Outreach*, 11(4), 1–17.

Coley, J.D. and Tanner, K. (2015) Relations between intuitive biological thinking and biological misconceptions in biology majors and nonmajors. *Life Sciences Education*, 14.

Colsoul, A., Carette, V., de Biseau D'Hauteville, J.-C., Poncelet, J.-F., Wolfs, J.L. (2009). Étude du processus d'appropriation de la notion d'évolution biologique par les élèves de l'enseignement primaire et secondaire en Communauté française de Belgique dans le but de la mise au point d'outils didactiques destinés à faciliter cet apprentissage. Final research report (unpublished), Université Libre de Bruxelles, Brussels.

Coupaud, M., Delserieys, A., Jégou, C., Brandt-Pomares, P. (2015). Pupil's conceptions of biological evolution throughout secondary school in France. In *ESERA 2015, Science Education Research: Engaging Learners for a Sustainable Future*, Helsinki.

Crépin, P. (2001). Les systèmes explicatifs des enfants de l'école primaire sur l'origine des espèces. Thesis, École Normale Supérieure de Cachan.

Crépin, P. (2002). Des conceptions initiales aux systèmes explicatifs des élèves de l'école primaire sur l'origine des espèces. *Grand N*, 70, 101–123.

Darwin, C. (1866). *On the Origin of Species*. John Murray, London [Online]. Available at: http://darwin-online.org.uk/content/frameset?viewtype=text&itemID=F385&pageseq=1.

Darwin, C. (1868). *The Variation of Animals and Plants Under Domestication*, 1st edition. John Murray, London.

Desmates, S.S., Settlage Jr., J., Good, R. (1995). Students' conceptions of natural selection and its role in evolution: Cases of replication and comparison. *Journal of Research in Science Teaching*, 32(5), 535–550.

Fabre, M. (2016). *Le sens du problème. Problématiser à l'école ?* De Boeck, Brussels.

Fabre, M. and Orange, C. (1997). Construction des problèmes et franchissement d'obstacles. *Aster*, 24, 37–57.

Fortin, C. (1993). L'évolution : du mot aux concepts. Étude épistémologique sur la construction des concepts évolutionnistes, et les difficultés d'une transposition didactique efficace. PhD Thesis, Université de Paris.

Fortin, C. (2009). *L'évolution à l'école, créationnisme contre darwinisme ?* Armand Colin, Paris.

Gayon, J. (1992). *Darwin et l'après-Darwin une histoire de l'hypothèse de la sélection naturelle*. Kimé, Paris.

Gayon, J. (1997). *Hasard et évolution*. Dossier Hors-Série, Pour la Science.

Gayon, J. (2006). Evolutionnisme (Biologie). In *Dictionnaire d'histoire et de philosophie des sciences*, Lecourt, D. (ed.). PUF, Paris.

Gobert, J. (2014). Processus d'enseignement-apprentissage de raisonnements néodarwiniens en classe de sciences de la Vie et de la Terre. PhD Thesis, Université de Caen Basse-Normandie.

Jégou, C. (2009). L'enseignement de l'évolution des espèces vivantes à l'école primaire française. Rapport au savoir d'enseignants et d'élèves de cycle 3. PhD Thesis, Université d'Aix Marseille I, Université de Provence.

Jimenez-Aleixandre, M.P. (1992). Thinking about theories or thinking with theories? A classroom study with natural selection. *International Journal of Science Education*, 14(1), 51–61.

Kettlewell, H.B.D. (1956) Further selection experiments on industrial melanism in the Lepidoptera. *Heredity*, 10, 287–301.

Kettlewell, H.B.D. (1959). Darwin's missing evidence. *Scientific American*, 200(3), 48–53.

Lhoste, Y. (2008). Problématisation, activités langagières et apprentissage dans les sciences de la Vie. PhD Thesis, Université de Nantes.

Lhoste, Y. and Gobert, J. (2022). Conquérir un point de vue populationnel et historique pour apprendre l'évolution. Analyse de la co-activité professeur-élèves par l'étude de la structuration des contextes. *Éducation & Didactique*.

Majerus, M.E.N. (1998). *Melanism: Evolution in Action*. Oxford University Press, Oxford.

Majerus, M.E.N. (2008). Industrial melanism in the peppered moth, Biston betularia: An excellent teaching example of darwinian evolution in action. *Evolution: Education and Outreach*, 2, 63–74.

Mayr, E. (1989). *Histoire de la biologie. Diversité, évolution et hérédité.* Fayard, Paris.

Metz, K.E., Sisk-Hilton, S., Berson, E., Ly, U. (2010). Scaffolding children's understanding of the fit between organisms and their environment in the context of the practices of science. *Learning in the Disciplines: Icls 2010 Conference Proceedings – 9th International Conference of the Learning Sciences*, 1, 396–403.

Ministère de la communauté française de Belgique (2001). Socles de compétences. Report, Ministère de la communauté française de Belgique.

Orange, C. (2000). Idées et raisons : construction de problèmes, débats et apprentissages scientifiques en sciences de la vie et de la Terre. Unpublished Education Sciences HDR Thesis, Université de Nantes.

Orange, C. (2003). Débat scientifique dans la classe, problématisation et argumentation : le cas d'un débat sur la nutrition au cours moyen. *Aster*, 37, 83–107.

Orange, C. (2010). Situations forcées, recherches didactiques et développement du métier enseignant. *Recherches en éduction*, 2, 73–85.

Orange, C. (2012). *Enseigner les sciences. Problèmes, débats et savoirs scientifiques en classe.* De Boeck, Brussels.

Orange Ravachol, D. (2005). Problématisation fonctionnaliste et problématisation historique en sciences de la terre chez les chercheurs et chez les lycéens. *Aster*, 40, 177–204.

Paley, W. (1809). *Théologie naturelle ou preuves de l'existence et des attributs de la divinité*, 12th edition. J. Faulder, London.

Peterfalvi, B. (2008). Comprendre la théorie de l'évolution, une course d'obstacles ? In *L'évolution du vivant : un enseignement à risques*, Coquidé, M. and Tirard, S. (eds). Adapt SNES – Vuibert, Paris.

Plantin, C. (1996). Le trilogue argumentatif. Présentation de modèle, analyse de cas. *Langue française*, 112, 9–30.

Rumelhard, G. (1986). *La génétique et ses représentations dans l'enseignement.* Peter Lang, Bern.

Toulmin, S. (1993). *Les usages de l'argumentation.* PUF, Paris.

Uyen, A.L. (2013). The nature and variability of children's alternative conceptions of evolution. PhD Thesis, University of California, Berkeley, CA.

12

Obstacles and Challenges in Teaching Probabilistic Population Thinking in Evolutionary Biology – A Case Study

Julie GOBERT[1] and Laurent THEIS[2]

[1] *ADEF, INSPE Aix-Marseille Université, France*

[2] *Université de Sherbrooke, Canada*

12.1. Introduction

The purpose of this chapter is to identify a number of difficulties or obstacles related to probabilistic thinking in the processes of teaching and learning concepts from evolutionary theory. Different meanings of chance are involved in evolutionary theory: it may refer to the randomness of genetic mutations, the contingency of evolution or the use of probabilistic models to understand the idea of evolutionary forces, including natural selection and genetic drift.

We wished to approach this subject by using an epistemological analysis. This is focused on the concept of randomness in the evolutionary sciences and in mathematics. In the logic of Darwinian reasoning, two forms of randomness are present: the randomness of the convergence of causally independent series that, once identified, becomes a historical fact and refers to the idea of contingency, and the randomness of the genetic variability inherent in living beings, which is a theoretical

The Evolution of Life: Teaching, Learning and Training,
coordinated by Corinne FORTIN and Julie GOBERT.

postulate in Darwin's problematization, for whom "all non-hereditary variation is of no interest" (Darwin 1989, p. 13). We then study the implementation by an Earth and Life Sciences (ELS) teacher of an interdisciplinary situation that proposes the introduction of a mathematical probabilistic model in the 3rd grade (14–15 years old) for work on the understanding of the concept of evolutionary forces, in this case natural selection and genetic drift.

This study also makes it possible to question the difficulty of working on interdisciplinary situations. It is indeed a question of succeeding in linking two disciplines that call upon different epistemologies, notably on the ideas of probability or, more broadly, of chance.

The use of probabilities in biology to produce explanations is plural. Problems in biology can be constructed within distinct and varied theoretical and methodological frameworks, which generate different types of explanations within this field of research (Morange 2005). These types of explanations are not necessarily opposed to each other, and can even be linked together. Two types of explanation interest us in terms of their use of probabilities. On the one hand, biology can develop explanations based on strict determinism: causal explanations are then based on systems of laws, linear causal explanations of a mechanistic type. There is then no room for chance in the phenomena studied, no random phenomena. Other types of explanations are also produced in biology, in particular those based on probabilistic modeling where the stochasticity of certain phenomena is considered. This stochasticity is modeled using the mathematical tools of probability. These are two distinct theoretical and methodological approaches where the use of probabilities is possible. However, their status differs, and this is what we will try to explain in order to shed light on possible epistemological obstacles with regard to what interests us, namely the teaching and learning of evolution.

These epistemological analyses will then help us analyze what is at stake in the interdisciplinary classroom situation on natural selection and genetic drift.

12.2. Epistemological anchoring

12.2.1. *Strict Laplacian determinism is incompatible with the construction of evolutionary explanations*

Astronomer and mathematician P.-S. Laplace (1749–1827) is one of the fathers of probability. In his book *Principes mathématiques de la philosophie naturelle* (1814), he reduced "all celestial phenomena to a single formula", the law of universal gravitation. In his *Essai philosophique sur les probabilités* (1814), a famous quotation allows us to identify the theses of strict Laplacian determinism:

> We must therefore consider the present state of the universe as the effect of its previous state, and as the cause of that which will follow. An intelligence which, for a given moment, would know all the forces of which nature is animated, and the respective situation of the beings which compose it, if moreover it were vast enough to submit its data to analysis, would embrace in the same formula the movements of the greatest bodies of the universe and those of the lightest atom: nothing would be uncertain for it, and the future as well as the past, would be present to its eyes (Laplace 1986).

What are these theses? First of all, any event or thing cannot occur without a cause; this is how the principle of sufficient reason is stated. It is then a matter for science to discover and establish these causes. This causality is here entirely mechanical. Moreover, the universe is considered a whole that can be known by analyzing its successive states. These two theses are part of an ideal of science characterized by its power of prediction and the certainty of its statements. In this context, what role do probabilities play for Laplace?

Laplace tells us that "the uncertainty of human knowledge concerns events or the causes of events". His theory of probability as he presents it allows us to work on two types of problems. He sees two interests in the calculation of probabilities:

– The possibility of predicting the realization of an event with certainty. He reasons from an urn of known composition (deduction).

– The possibility of tracing events to their causes. He reasons from an urn whose composition is sought through samples (induction).

Laplace, in the search for causes, wants to establish certainties: a probability of 0 or 1. The general law of probabilities then makes it possible to look at how we deviate from the "truth".

According to Anne Fargot-Largeault (2009), within the framework of strict determinism, probabilistic analysis postulates the absolute invariability of causes, which amounts to thinking that the distribution of possibilities of events is established once and for all by the laws of the universe. On the contrary, if these distributions can be subject to random fluctuations, we can conceive of a history and thus an irreversibility, an evolution. The problems of evolutionary biology then call for a time constructor of phenomena (Orange Ravachol 2012). To think of stochasticity is to think that there can be a random distribution (a realization) of events, and to integrate this dimension into a model of the phenomena. However, in the context of strict Laplacian determinism, these events must be linked to constant causes. For Laplace, probabilities are a means of establishing certainties, universal laws that apply to any object through the observation of regularities. However, neo-Darwinian explanations correspond to a form of explanation made necessary by the existence of a historical dimension in the field of life sciences. In evolutionary biology, the field of relevance of neo-Darwinian probabilistic modeling is that of the study of complex systems, the evolution of which we wish to understand over time, and which are therefore bearers of a historicity, the objects of which are populations of discontinuous and independent elements, and the explanations of which couple randomness with selection.

12.2.2. *Use of probabilities in evolutionary concepts*

12.2.2.1. *The forms of randomness in the theory of evolution*

Several definitions of randomness are at play in evolutionism. In a common sense, it means luck and bad luck. For example: "the gardener digging up his vegetable garden to plant seeds finds a gold coin" (Gayon 1994). In evolution, the randomness of mutations is also a question of luck or misfortune: a mutation does not occur because it is useful, but it can have consequences, important or not, for the organism and its descendants (ibid.). "Randomness", used in this way, is focused on the idea that variation is random (random mutations) and is not "intrinsically directed toward adaptation, not that all mutational changes are equally likely" (Gould 2006, p. 11451). This "randomness" envisioned by evolutionary explanations implies an absence of finalism in all variation.

Randomness also refers to random events, the determining conditions of which we do not know. Throws of a six-sided die thus correspond to a series of random events. To know which of the six possibilities will come true, we would need to know all the parameters of the throw, but we are unable to determine these conditions. To control this randomness, we use the calculation of probabilities. A random event is then an event that can be interpreted by a law of probabilities. This recourse to probabilities, which concerns the prediction of random events, is found in the concept of reproduction and in the idea of evolutionary forces in population genetics, to work on natural selection and genetic drift in particular. We will develop this below.

The idea of randomness-contingency, finally, is also conceived as fortuitous, contingent events, in a historical perspective. The idea of filiation conceives living beings as changing and part of a historicism. Their study thus requires a particular approach that is aimed at retracing the history of the evolutionary process of the populations of living beings. In a Darwinian explanation, Fagot-Largeault (2009, p. 91) tells us that "what we are looking for to explain the change, it is thus a singularity valued by the chance of a fortuitous meeting". We therefore seek to identify events and trace chronologies and explain the causes. We seek to reconstitute a chain of events that are considered causal. The explanations are retrospective in nature.

12.2.2.2. *Mendel and reproduction*

The foundations of the intelligibility of neo-Darwinism are largely to be found in the science of heredity, i.e. Mendelism. This structures and organizes the field of scientific knowledge related to questions of genetics and evolution (Gobert 2020) and it is in this sense that we call it a structuring principle, in reference to the work of Denise Orange and Françoise Beorchia (2011). At the beginning of the 20th century, the articulation between the contributions of Mendel and neo-Darwinism made it possible to switch to population genetics. This is a condition of the possibility of the neo-Darwinian theory of evolution. "Current genetics, and particularly population genetics, integrates the Darwinian concept of selection, the Mendelian rules of transmission and the Mendelian–Darwinian patterns of an organism now represented less as a functional solidarity than as a combination of particles at the level of which the behavior of the individual and the fate of the species are played out" (Conry 1987, p. 124) and "by clarifying the experimental meaning of concepts such as variation and heredity, Mendelian genetics transformed what was in Darwin only a pre-requisite of the principle of selection into an experimental and theoretical basis on which it was possible to reconstruct the concept of selection" (Gayon 1992, p. 261).

Mendel is interested in the problem of the mechanisms of heredity, i.e. the mechanisms explaining the modalities of the transmission of phenotypic characters of individuals. His work is indifferent to the problem of transformism. Mendel's work (1865) is based on a probabilistic model that uses Bernoulli's mathematical model of random draws with replacement. His point of view leads him to be interested in the phenotypic characters of the individuals of the second generation (F2), since it is in F2 that we can again find phenotypic characters that were also observable in the parents. The interpretation that he will build of the forms observed in F2 is that of a model of rigorous mathematical distributions resulting from probabilistic combinatorics. The phenotypic characters are separated according to the proportion 1+1+2 (AA + aa + 2Aa), which can be represented by a Punett square method. He links his modeling work to experimental testing work at the population scale.

In Mendel's work, the design of his experimental plans is guided by the law of large numbers and probabilities, which runs counter to the naturalist attitude that always puts observation first (Rumelhard 1986). To build his model, he imagines mutually independent particles that can separate and recombine. He thus overcomes the idea-obstacle of heredity as a mix. He is familiar with the work of Laplace and introduces randomness to the formation of gametes and fertilizations. Implicitly, he imagines an urn of gametes and a "random" draw with a replacement and simulates the expected results (Mendel 1961). He was expecting theoretical results that he wanted to compare to observed results. He even challenged some of his own results and explicitly repeated some experiments. Hence, the false accusation of "fraud" (Rumelhard 1980) is, on the contrary, the best proof of the fact that he expects a result and does not go back inductively from the empirical to the theoretical results. However, he does not have tests to assess the discrepancies. These would have made it possible to know that we can also reason about small numbers under certain conditions. Statistics and, above all, comparison tests would come 40 years later at the beginning of the 20th century. With this theoretical choice, Mendel admits that it is sufficient to know the "n-1" generation to statistically predict the "n" generation. The idea of discontinuity and the independence of what would later be called "genes" is a condition of possibility of Mendel's work. This concept of particles also rejects any idea of heredity by mixing characters, any hybridization in the sense of a fusion of characters. A "black box" reasoning (Jacob 1970) makes it possible to consider only the inputs (supposed gametes) and outputs (phenotypes of flowers and fruits) while ignoring the intimate mechanisms of this transmission. Theoretical logic would have dictated that we begin with the chemical nature of the "genes", their location and the mechanisms by which they control the realization of the characters. It is the formidable efficiency of mathematics to be able to bypass all these questions that seem a priori to be unavoidable conditions of possibility. Mendel models the transmission of hereditary characters by constructing a new and original world through the mathematical treatment of the phenomena he proposes: "the Mendelian factor, a conceptual object, is at first virtual, like a mathematical variable. Suddenly, two worlds appear: a mathematized world, virtual, complex to reach, but where some simple rules underlie the algebra of the factors; another world, visible, or rather apparent, that can be described by our senses, and whose apparent disorder is totally dependent on the ballet of the 'factors' of the first world" (Leguyader 2012, p. 148). In this, he invents a new discourse. With Mendel's work, a rule for constructing the transmission of characters is now conceived and opens up a new conceptual field. The principle of the segregation of characters organizes the new logic of heredity around two fundamental schemes: a particulate (or discrete in the mathematical sense) representation of the organization, correlative to the affirmation of a random combination of these characters. According to Yvette Conry (1987, p. 129), in order to allow "the transformation of the theoretical support and the conceptual field" that Mendel achieves, a method, a mathematical treatment of the data and a symbolic representation were necessary. His work is based on a probabilistic mathematical modeling of large numbers and the idea of sampling; he invents and uses a new mathematical symbolization, takes into account a population of discrete elements and then considers the meeting of gametes as random combinations of these discrete elements. It took 35 years for this innovative work and this use of mathematics in biology to be accepted and pursued. The existence of random phenomena or events that are essentially unpredictable has long been explored by biology, and here the randomness of encounters and combinations of "genes" means that not all combinations occur, even if we can statistically predict the probabilities of each one; we cannot predict the combinations that will actually occur (no predictable law in the sense of strict determinism). This makes it possible to conceive contingency and historicity in the transmission of characters over the generations.

12.2.2.3. *Hardy–Weinberg model in population genetics*

The articulation of Darwinism and Mendelism in 1915, with Morgan and Punnet, allowed the appearance of the new scientific field of population genetics.

Population genetics is a field of biology that is concerned with the evolution of the genetic composition of populations within a species. It seeks to understand the short-term and long-term evolution in a population, i.e. the "unique and irreversible" process of biological evolution (Goux 1979). The articulation of Mendelism and Darwinism allows for a change in the level of organization that characterizes the epistemological framework of population genetics. The reasoning is at the level of genes and alleles, in terms of frequency evolution. The probabilistic description of the combinations of genes and alleles at the two stages of sexual reproduction (fertilization, meiosis) has introduced new modalities of evolutionary changes and has opened the door to the new methodological and theoretical field of population genetics. We can refer to mathematical theory because "the models of population genetics are essentially probabilistic models" and "there are few fields of biology where the modelling activity has been as extensive as in population genetics" (Goux 1979, p. 565). The overall population size has emerged as essential to the treatment of problems, as well as to the real future of the genetic composition of the population. Depending on whether we consider populations to be infinite or finite, the assumptions differ and so do the problems.

The work of Fisher, Wright and Haldane, for whom the elementary unit in evolution is the gene, taken as a unit of mutation, function and segregation, constitutes an essential contribution. The Mendelian theory, often called Mendel's laws, was able to account for the hereditary determinism of the evolutionary process proposed by Darwin. This period founded the synthetic theory of evolution or neo-Darwinism. The Hardy–Weinberg equilibrium model deals with the fate of allelic frequencies and genotypic frequencies in a population of diploid individuals, assumed to be infinite, panmictic (crossings between individuals are random), without mutation, migration or selection. Modeling, under these conditions, of the genetic transformations of populations, resulting from the successive meiosis and fertilization of sexual reproduction, reveals a persistent equilibrium despite the constant reshuffling of the allelic associations in the genotypes; alleles that are nevertheless dissociated by meiosis at each generation (Goux 1979, p. 567). Population genetics is not interested in populations as collections of adult individuals characterized by their phenotypes and genotypes, but will consider gamete populations. This makes it possible to rely on a probabilistic model: the gametes, as independent and identically distributed elements, constitute at each generation an "urn of gametes" modeled by a "Bernoulli urn". Under the conditions stated above, the Hardy–Weinberg model tells us that if we consider a single locus, the frequencies of the A and a alleles of a gene of the species considered within this population being p and 1-p = q, they do not vary from one generation to the next. The proportions of the genotypes at the second generation and for all other generations are: $F(AA) = p^2$, $F(Aa) = 2pq$ and $F(aa) = q^2$. They remain stable over time; they are at "equilibrium". These formulas are established from the Mendelian rule of transmission of characters during sexual reproduction in a situation of panmixy. The stability of the population is the necessary consequence and means, in other words, that allelic variability never disappears. "Mendelian rationality leads us to posit the indefinite persistence of residual variability. This is the conceptual root of the Hardy–Weinberg law" (Gayon 1992, p. 302). It is the reference model on which population genetics is based (Petit and Zuckerkandle 1976, p. 62); it is its starting point (Gayon 1992, p. 300). The explanations of evolution by selection by population geneticists are thus always linked, even implicitly, to this model (Sober 1984). Proposed in 1908, it fits in well with a series of research projects that rediscover Mendel's work and where Mendelism and Darwinism meet. The theoretical work generated by this model will have allowed unification of the explanations of biological evolution.

If one of the conditions of the Hardy–Weinberg model is not met, then the evolution of allele frequencies is different from that predicted by the model (Huneman 2009). Researchers thus establish "selective values" for individual genotypes. This means making the hypothesis of their more or less advantageous character within a population. In turn, the hypothesis on the selective value of different genotypes allows predictions on the deviations of genotype frequencies in the population with respect to the Hardy–Weinberg equilibrium. These predictions are based, in this mode of reasoning, on the hypothesis of an infinite population. A series of conceptual breaks within evolutionary theory are thus made possible. Jean Gayon (1992, p. 304) summarizes them in part as follows:

> The Hardy–Weinberg model provides a reference for evaluating, quantifying, and detecting the various effects of evolutionary forces, and makes mutations and selection appear not as alternative explanations of each other, but "as interacting forces in the dynamics of gene frequencies" (Gayon 1992, p. 304).
>
> The panmictic model envelops the idea of the Mendelian population as a "gene pool", to which all individuals contribute equally. Thus "evolutionary forces" can be defined by their effects "on the homogeneous field of gene and genotype frequencies" and "each factor can be described in its effect as a 'deviation from panmixy'" (Ibid.).
>
> The statistics of population genetics are given as a prediction of the future of a population from an initial state [...] they have the appearance of kinetics which predict the rate of change under such and such a hypothesis. It is therefore from the outset genuinely experimental in its project, since it is based on causal hypotheses and is exposed to the sanction of experiments (Ibid.).

The concept of selection seen as differential reproductive capacity has undergone an essential transformation: selection is not only an evolutionary process that is played out in the selective value of individual phenotypic traits, but it is now seen as kinetic and capable of predicting the speed of diffusion of an allele in a population. It is now linked to the idea of mutation, which generates new alleles. Selection can no longer be understood "as the selection of the best".

The overall size of the population very quickly appeared to be essential to the treatment of the problems of population evolution. The phenomenon of genetic drift, and in particular its importance as an evolutionary process, was the subject of lively controversy, notably between two population geneticists: Fischer and Wright, in the 1920s and 1930s, and then in the second half of the 20th century.

Sewall Wright (1889–1988) asked himself the following question: how can we envisage the evolution of genotype frequencies in a finite population? The Hardy–Weinberg model was used to model the evolution of an infinite population and we have seen above that the combined play of selection and mutations tends to fix gene frequencies on equilibrium positions. This is a prediction from the model. Even under conditions assumed to be perfectly constant, these equilibria are only really maintained without change from one generation to the next if the population size is extremely large, unlimited in principle. However, Wright was particularly interested in small populations: on the one hand, the empirical data contradicted what the Hardy–Weinberg model predicted, and, on the other hand, "large allelic fluctuations" were indeed observed. The discrepancy between predictions and empirical data will require a rethinking of the model of allelic frequency evolution. To demonstrate the importance, the role, the influence of the population size in its evolutionary possibilities, new hypotheses will be necessary and new models and mathematical tools will also be needed. Wright uses a new probabilistic mathematical model, which is different from the Hardy–Weinberg equilibrium. This probabilistic modeling of the evolution of allelic and genotypic frequencies in a finite population will allow him to conceptualize, as early as 1917, a major phenomenon for evolution called random genetic drift. The explanatory hypothesis is the following: the limitation of the number of reproducers always introduces, in fact, a tendency towards a random variation of gene frequencies during the succession of generations or stochastic fluctuations. Given a parental generation with known genetic characteristics, the allele frequencies in the population of the next generation are not rigorously determined, but given by a probability law.

This random deviation can lead to different paths in the evolution of allelic frequencies according to the populations. The field of possibilities is very broad and cannot be explained by a causal determinism, but is instead probabilistic. The evolutionary trend is modified according to the case. In probabilistic models, there is a causal indifference regarding the nature of a character in the explanation of its frequency variations. The theory of biological evolution has a stochastic character, in that it is a theory that "from the same set of initial conditions, it is possible to predict, not one and the same result, but several possible results according to a certain law of probability" (Malaterre and Merlin 2009, pp. 337–338). While statistical predictions about the evolution of allelic frequencies are possible, nothing can predict the fate of a particular individual. If selection and genetic drift are both probabilistic sampling processes, selection is defined as a probabilistic sampling process that discriminates on the basis of differences between organisms. Genetic drift is defined as an indiscriminate probabilistic sampling process (unrelated to the traits of individuals in interaction with the environment) (Beatty 1984; Hodge 1987). These historical and epistemological analyses allow us to say that the gamete urn model and sampling processes are the foundation of the explanations and reasoning produced by population genetics.

The epistemological analysis carried out would require further development and is still limited here, but it allows us to highlight the objects of knowledge involved in the teaching sequence which is the subject of our corpus. Reflections such as those carried out by Walsh et al. (2002), which distinguish between dynamic and statistical interpretations of the theory of evolution, are also quite enlightening in showing the importance of a statistical interpretation for understanding the selection/genetic drift couple.

12.2.3. *Probability: the specificities of probabilistic thinking and the approaches used in the mathematics classroom*

In this section, our aim is to draw a general picture of the characteristics of probabilistic thinking as it is generally conceptualized in mathematics education. We will also discuss two of the main approaches that are used to deal with probabilistic situations. This contextualization will allow us to situate the epistemological obstacles linked to the use of probabilities in mathematics and evolutionary sciences, which will be at the center of our analyses.

First of all, within mathematics, probabilities occupy, by their nature, a particular place. Indeed, in other fields such as arithmetic and algebra, the same operation, performed under the same conditions, will always generate the same result. We can then approach these domains with a deterministic logic. However, this is not necessarily the case in the field of probability, since two experiments, even if carried out under similar conditions, can generate different results. This, then, is a different logic, which requires us to think in a non-deterministic way in situations that involve chance.

From a strictly mathematical point of view, there are generally three approaches to determining a probabilistic value. Here, we will study two of them in depth, namely the theoretical approach and the frequency approach[1].

In a first approach, called the theoretical or cardinalist approach, the probabilities are obtained using rigorous reasoning or a formula. Generally speaking, the probabilities obtained are then expressed as the ratio between the number of favorable cases and the number of possible cases. This approach goes back to Laplace: "The theory of probability consists in reducing all the events that can take place in a given circumstance to a certain number of equally possible cases, [...] and in determining among these cases the number of those that are favorable to the

1. The third approach, the subjective approach, is based on an intuitive evaluation of a probability. As it is rarely explored in a school context, we will not explore it further here.

event, whose probability is sought" (Laplace 1812, cited in Parzysz 2019, p. 48). This approach is then entirely in the realm of probability theory and does not involve experimentation in the real world (idem, p. 53).

The so-called frequency approach is based on the principle of carrying out successive tests under the same conditions. As random phenomena are involved, the results obtained are necessarily variable. The probabilities obtained in a frequency approach cannot then be approached solely in a deterministic way. Moreover, the frequency approach allows us to obtain "an estimate of the probability of the event, obtained on the basis of the observation of the frequency observed after repetitions of the experiment" (Parzysz 2019, p. 53). As a result, unlike the theoretical approach, the frequency approach is primarily rooted in experiments, in the real situation.

It is the law of large numbers that stipulates that the greater the number of trials carried out (frequency approach), the closer the results tend to be to the theoretical probabilities, which constitutes the bridge between the two approaches. In a teaching context, the articulation of these two approaches is not easy as it requires a real change of posture to move from one approach to the other. Moreover, the teaching of probability is often considered by mathematics teachers to be rule-centered teaching, in a theoretical and deterministic approach (Stohl 2005, cited in Martin and Theis 2016, p. 345). Moreover, in the theoretical approach, the same situation under the same conditions always leads to the same theoretical probability, which in a way allows a deterministic logic to be maintained there, whereas the frequency approach implies a sampling variability. As a result, when a small number of trials are carried out in an experiment, the results obtained in a frequency approach do not necessarily correspond to the results expected in a theoretical approach and it becomes difficult for the teacher to manage these differences. Moreover, the research we have conducted with elementary school teachers (Martin and Theis 2016; Theis et al. 2019) to develop situations that enable the articulation of the frequency and theoretical approaches shows the difficulty, in terms of teaching, of this link. In a first experimentation that was aimed at linking the theoretical and frequency approaches, it was difficult for the teacher to leave behind a theoretical approach and a deterministic perspective (Martin and Theis 2016). A second experimentation, which started from a frequency approach to a situation and then bridged to theoretical probabilities (Theis et al. 2019), showed the difficulty for students in anchoring their experiments and the results obtained to a theoretical approach.

12.2.4. *Some ideas to consider the difficulties of linking probability with the teaching of evolution in the ELS class*

At the epistemological level, several fundamental differences between the use of probability in mathematics and science can make it difficult to articulate them in a science course, and particularly in a situation dealing with evolution.

The first of these differences concerns the sample space, i.e. the set of possible events in a probabilistic experiment. As a result, in a classical probabilistic situation, this space is invariant for a given situation. For example, the sample space of the throw of a regular die is {1, 2, 3, 4, 5, 6} and remains stable from one roll to the next. By repeating the experiment of throwing a die, even a large number of times, this space remains invariant and no possibilities other than the ones initially expected can appear. The situation is somewhat different when these concepts are applied to the transmission of characteristics from one generation to another. Indeed, genetic mutations, occurring at random, can appear and generate the appearance of new cases that were not foreseen (and were not foreseeable before) and new combinations of alleles, during reproduction, can also occur; this also contributes to variability within populations. The sample space can therefore change as a result of genetic mutations and reproductive mixing.

The non-distinction between a sample space that is not changeable in a mathematical situation, and the possibility of new cases appearing in the living and thus a sample space that is constantly changing in biology could be a rather strong obstacle for students, especially if it remains implicit.

A second difference concerns the stability of the experimental set-up, i.e. the urn in which the draws are made. In classical probability situations, the different trials are always carried out from the same starting device. For example, to determine the probability of drawing a white ball from a box containing five white balls and five black balls, the theoretical approach would involve a ratio between the number of favorable cases and the number of possible cases. The frequency approach would involve drawing a ball from the box a large number of times (with a discount) and compiling the results. However, to do this, the initial urn of five white balls and five black balls must remain constant and cannot be modified from one draw to the next. This is not the same situation when we study the reproduction and transmission of characteristics from one generation to the next. Indeed, different factors can contribute to the fact that, from one generation to the next, the starting composition is not the same. The urn of gametes used in the biology course corresponds in a certain way to the drawing urn of the probabilistic situation. However, its composition is likely to change from one generation to the next and this change remains implicit for the students. On the one hand, simple sampling variation, what we call randomness, can cause a given characteristic to appear more frequently from one generation to the next than would be theoretically expected and thus change the distribution of characteristics for the next generation. On the other hand, environmental conditions may also cause individuals with certain characteristics to have a better chance of reproducing (selective value effect). In the next generation, the individuals that are able to reproduce thus do not necessarily carry, in the same proportions, the characteristics of the previous generation. In a way, there can therefore be evolution from one generation to another in situations of reproduction, whereas evolution does not occur in purely probabilistic situations such as those studied in a mathematics course in secondary school.

The historical perspective is a point to keep in mind when using the probability model. Students need to be able to consider it in order to understand that the gamete urn is never the same over generations and that there may be "generational gene flow" that changes or does not change the allelic frequency distributions. As Jacob puts it:

> For a population to evolve, it is sufficient that the genetic background common to that population varies, either abruptly or gradually over generations. Such statistical variation in the relative survival of individual genes does not necessarily imply adaptation. It may simply reflect the effects of chance on some stage of reproduction (Jacob 1981, p. 43).

Students should then be able to trace the possible or necessary conditions for population transformation (mutation, migration, selection, drift), i.e. the reasons for changes in the composition of the urn from one generation to the next.

Mathematics and ELS have different epistemologies when it comes to mobilizing probabilities. This raises questions about possible difficulties in the implementation of interdisciplinary teaching–learning situations by ELS teachers. In the rest of this study, we will illustrate some of the difficulties that arise for both the teachers and the students at the interface of these two epistemologies.

12.3. Background to data collection and methodology

The corpus of data consists of a teaching–learning sequence on evolutionary forces (selection, genetic drift, migration, mutation) in the 4th year of high school

(15 year old students) in an explanatory register that would be neo-Darwinian as it articulates population genetics and evolutionism. This is a sequence that was implemented in June 2013, by an ELS teacher, in the framework of a research involving the same teacher and a researcher. The collaboration between the teacher and the researcher focused on a reflexive work on the conditions of didactic possibilities of construction of an evolutionary thought by students in the 4th year of high school, integrating the knowledge of genetics worked at this level of class. This reflection was based on an analysis of the socio-epistemological context of the knowledge at stake (Gobert 2020), together with its appropriation by the teacher. The progression of the sequence and the didactic situations were co-constructed by the teacher and the researcher, but the supports and associated written traces were then taken on and implemented by the teacher. There was no collaborative work between the sessions carried out in this sequence. The students in the class had already worked on genetics earlier in the year, which seems to us necessary for approaching evolution from a neo-Darwinian perspective with the students. Studies show the importance of working on genetic concepts beforehand (hereditary and non-hereditary traits, genes and alleles, homozygosity/heterozygosity, dominance and recessivity), particularly by using examples related to the human species (Banet and Ayuso 2003), which was the case for these students.

	Course of the sequence
Session 0	Diagnostic assessments
Session 1	Group work aimed at trying to link the genetic theory of heredity and the Lamarckian theory of evolution in order to identify the impossibility of the heredity of acquired characters
Session 2	Group work based on a problem situation concerning the study of the evolution of moth populations (*Biston betularia*) Discussion/remobilization of the Mendelian principle of gamete transmission and introduction of the Hardy–Weinberg model into the environment *(Video recording of the session and transcript of the discussion: the corpus includes 133 exchanges and is the subject of this chapter's analysis).* Group work aimed at producing an explanation of the evolution of moth populations different from a Lamarckian explanation
Session 3	Collective validation/structuring Individual writing of an explanation on the example of the evolution of giraffe populations (situation from session 1)

Table 12.1. *Course of the sequence (corpus used in 2013)*

In this sequence, the teacher tried to get the students to enter into probabilistic population reasoning. They needed to construct neo-Darwinian explanations of the evolution of populations that were focused on the conditions that make the evolution of allelic frequency distributions possible within populations over successive generations. More specifically, in this chapter, we attempt to look at how the teacher introduced a mathematical probabilistic model, the Hardy–Weinberg model, to understand the evolutionary forces at play in the evolution of a population. The teacher's intention here was to work on the deviation from the model, this deviation then allowing hypotheses to be made about the causes of the evolution.

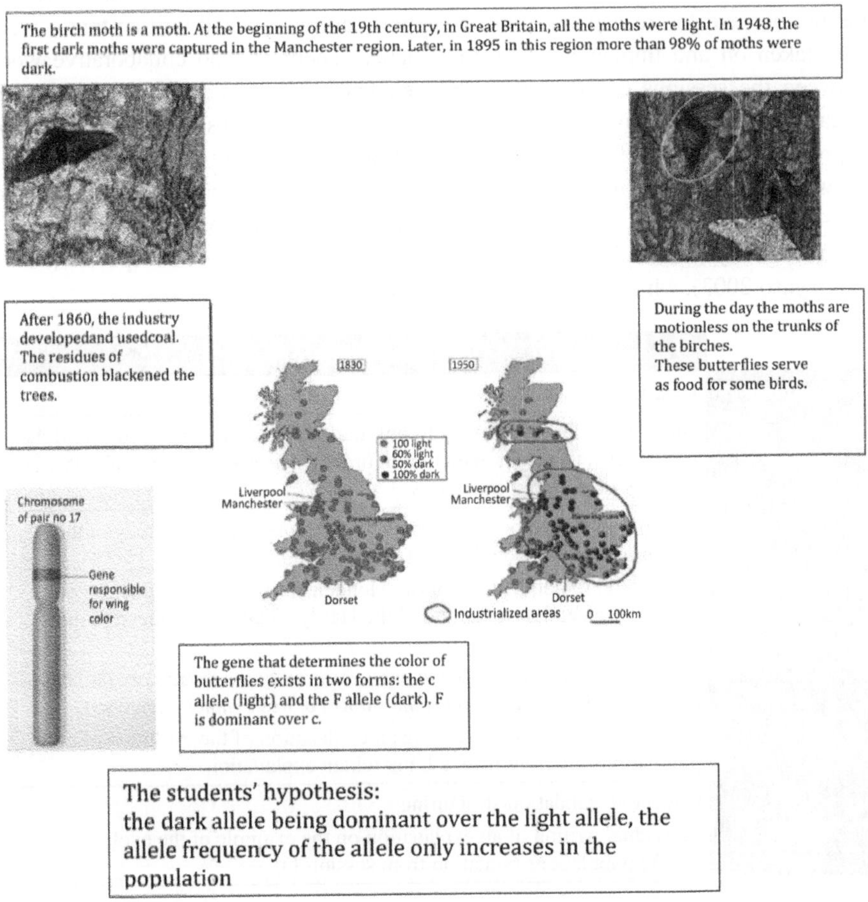

Figure 12.1. *Problem situation on the evolution of moth populations and students' evolutionary hypothesis*

The data analyzed are written records produced during the second session of the sequence implemented by the teacher (see Table 12.1) and the transcript of the group discussion recorded by video during this same session.

After a second reference problem situation given at the beginning of session 2 (see Figure 12.1), students explained the increase in the frequency of the "dark" allele in moth populations by making the following hypothesis: "since the dark allele is dominant over the light allele, the allelic frequency of the allele only increases in the population".

12.4. Analysis of teacher and student activity during group discussion

We proceeded to the thematic division of the collective discussion and presented a synthesis of what was said for each episode (Table 12.2).

	Topics covered
Episode 1 (1–10)	Presentation of the example of butterflies whose population is composed of an urn of gametes composed of 80% dark alleles and 20% light alleles
Episode 2 (16–20)	Students make hypothetical predictions about the evolution of the moth population → prediction: increase in the frequency of the black allele
Episode 3 (20–50)	Mobilization of a Punnett square to predict the evolution of allelic frequencies
Episode 4 (50–133)	Discussions on possible reasons for the evolution of allele frequencies in the moth population

Table 12.2. *Thematic breakdown of the discussion*

12.4.1. *Students' initial predictions about the evolution of the moth population (*Biston betularia*)*

The question of the "dominance" of the black allele, in the sense of "dominance versus recessivity" relative to the expression of genetic information and/or dominance in the sense of "being more numerous than…", intervenes explicitly or implicitly at different moments of the situation and seems to be an obstacle.

12.4.1.1. *Dominance versus recessiveness*

In episode 2, the teacher underlined the recessive character of the light allele ("here for example there is a light allele whereas the moth is dark"), which made it possible to make the link between the genotype and the phenotype of the individuals. They then presented the data of the initial allelic population: "so, if we count the small circles, we find 20% light alleles for 80% dark alleles/So here is a large population of moths/I drew a certain number, but we imagine a very large population". The teacher then asked about the possible evolution of this population: "In your opinion, if this population reproduces, will it evolve?" The students made a first hypothesis and here is what they thought: the black allele being dominant, its frequency only increased in the population during successive generations. Their hypothesis was accepted by the teacher and noted on the board. Here, they used the idea of allelic dominance, which refers more to the problem of the genetic expression of alleles. However, Hardy says:

> In a word, there is not the slightest basis for the idea that a dominant trait should show a tendency to spread throughout the population, or that a recessive trait should disappear (Hardy quoted in Gayon 1992, p. 303).

This idea-obstacle, which we see appearing here in the students' comments, is rooted in the common thought that those who dominate end up being more numerous or that the "strongest" resist better, while the "weakest" disappear. It is interesting to underline that the students did not consider a stability in time of the moth population at all: the situation proposed by the teacher indeed concerns a population that evolves. They foresaw a disappearance of the light alleles: "well, in many years there will only be dark alleles", and this is done with a long timeframe in mind: "in a very long time". The repetition of this idea and the reformulation of this idea by the teacher made it possible to explain the role of successive reproductions: "in a very, very long time/So we can see that the advantage of working with insects is that it goes faster in terms of generations/They reproduce much more quickly/Reproduction cycles are shorter"; it is not so much the time that is important as the number of generations. The teacher then rephrased the students' proposal as follows: "So you're telling me that if this population reproduces little by little over several reproduction cycles, the clear allele will disappear." As this hypothesis was shared and stabilized in the class, they then proposed to discuss and refute it by introducing the Hardy–Weinberg model via a Punnett square (see Figures 12.2 and 12.3). The teacher said: "and in fact, we can try to do it since we know how to make genetics work/How can I make a diagram of the reproduction of this population?" It seems interesting to say that at this point in the sequence, the students had not yet constructed the idea of selection or even differential predation, which can be considered as a first step in the elaboration of the concept of selection (Gobert and Lhoste 2021). It is therefore possible that the word "dominant", through the common meaning it conveys, may have had an explanatory value for the students (especially since the situation shows an increase in the frequency of the dominant allele).

12.4.1.2. *Dominance as "more numerous"*

Later in the discussion, the students explained this hypothetical evolution of the moth population through an analogy (which they believe to be true) with the evolution of "hair color" phenotypes in the human population:

> Student: "Well, in many years, there will only be dark alleles/well, yes, yes, because it's like blond and red hair, it's disappearing/there are more and more browns."
>
> Teacher: "Do you think at some point there will be no more blonds at all?"
>
> Student: "Well, in a long, long time."

It seems that because there are more brown-haired people, they will end up being the only phenotypic possibilities in the population, while blonds will disappear. We find this obstacle in episode 4, in which the students return to their hypothesis and tried to justify it: "but no, but because there is more chance of drawing a black one/because it is 80%". Here, since at the beginning the frequency of black alleles is more important, the evolution of the allele population will tend towards the disappearance of the light allele.

For the students, it seems to us that the reasoning was as follows: at the beginning, the black alleles have a greater probability of being "drawn", i.e. of participating in reproduction, and for this reason, as the generations go by, their frequency will increase all the more (there will be only dark alleles and no more light alleles at all). The most frequent at the beginning will be increasingly frequent, as it is the fact of being frequent in the population that advantages the allele. The most numerous will become increasingly numerous. This conception-obstacle could have been reinforced by the fact that the example chosen and presented to the students describes an increase in the frequency of the black allele, which, at the beginning, is already more frequent in the population and is dominant.

Moreover, it was probably difficult for the students to consider successive urns over several generations. Indeed, contrary to a simple probabilistic experiment, which always starts from the same starting situation, under the same conditions, it is necessary here to consider draws from several successive generations. The use of probabilistic simulations could be very useful in this context to model the evolution of populations through several generations. This would bring them back to a probabilistic reasoning; they could observe over several generations the composition of successive generations, and observe in particular that populations can remain relatively stable. This is the idea that the teacher wanted to introduce via the probabilistic Hardy–Weinberg model.

12.4.2. *Teacher's aims to bring out the ideas of survival and differential predation*

During the group discussion, the teacher introduced the Hardy–Weinberg model, without thematizing it as such, and used for this purpose a Punnett square, as well as an iconographic representation of a gamete urn (see Figure 12.2) where the gametes were represented by circles (spermatozoon or ovum); the black circles meant that the gamete had a "black" allele, the white circles a "white" allele. The triangles corresponded to the butterflies in the population. The phenotypic proportion in the population was indicated by a diagram. This document was completed collectively during the discussion on the class whiteboard via an overhead projector. These tools developed by the teacher operationalize the work relating to probabilistic population reasoning at the allelic level. This allowed the teacher to mobilize with the students the principle of gamete transmission, aiming at the joint management of the problem of heredity and the problem of the conditions necessary for the evolution of allelic frequencies in the moth population. The teacher, in her didactic use of the model, gave a refuting function to the Hardy–Weinberg model in the reasoning that they wished to have the students construct. Indeed, the Hardy–Weinberg model shows a stability of the evolution of the allele frequencies over time under certain conditions, and this is what they put forward to refute the first hypothesis of the students (who thought that the white allele would eventually disappear in favor of the black allele). They spoke about the large size of the allele population and emphasized the idea that "all gametes in the urn participate in reproduction". They did not talk about the following conditions of the Hardy–Weinberg model: no selection, no mutations, no migrations, as that is what they wanted to be able to work out with the students. They then gave the students the approach they wanted them to take in order to solve the problem:

> So, it is possible that by chance all the gametes that are there do not participate in the production of the next generation? We have to imagine a kind of big container in which we can draw.

As a result, for the teacher, the problem to be solved seems to be to explain why all the gametes in the urn did not participate in reproduction. If not everything is "picked" at each generation (this is how the teacher sometimes expressed herself), we must look for biological or ecological reasons why certain alleles are not "picked":

Does randomness play a role in reproduction? Well, yes, it does, it does at several levels. You agree that in order for me to show you that the population does not evolve, I have considered that all butterflies produce gametes and that all gametes produce young. Is it really like that?

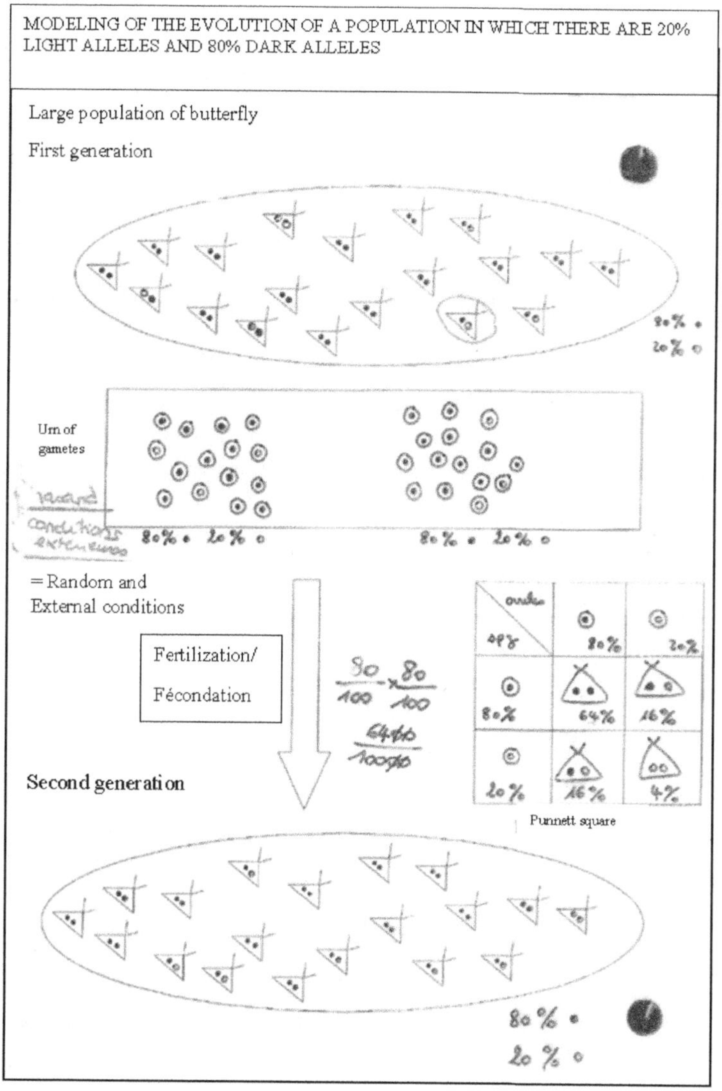

Figure 12.2. *Handout completed by the teacher and students overlaid on the board*

It seems to us that the teacher's intention here was to question the students about the reasons, in a biological or ecological context, as to why not all gametes participate in reproduction. There is thus an important shift: here, two phenomena seem to be superimposed, that of differential reproduction in butterflies and that of random fertilization at the origin of new individuals. This led the class to consider the phenomena of differential survival and differential predation at the level of the gametes produced by the individuals. The students returned to an example that they had studied a few weeks earlier in a course on reproduction; the case of sea urchins that produce numerous gametes in water: "we had seen, there are water currents that carry away gametes, predators that eat them". This created an ambiguity that could hinder the construction of the idea of selection and the understanding of the evolution of allelic frequencies in a population.

12.4.3. *A presentation of the model from a deterministic perspective*

During reproduction, the allelic frequencies in the population of the next generation are not rigorously determined, but given by a probability law. Instead, the teacher started from the theoretical model, and then deduced a distribution in the next generation that would be identical to the initial one, in a deterministic way, without taking into account the sampling variability. This is what we find, for example, in the following extract: "the second generation is exactly the same as the first, so there is no evolution".

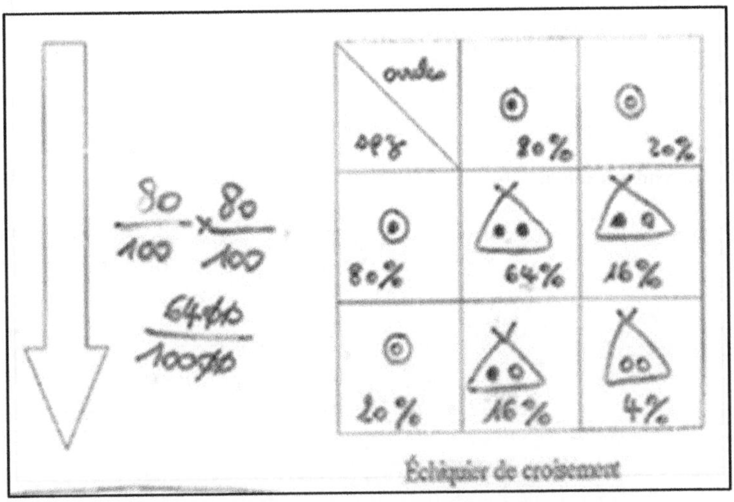

Figure 12.3. *Details of Figure 12.2: collective paper trail following the use of the Punnett square*

During the experiment, when the teacher presented the Punnett square, which made it possible to designate the possible phenotypes and genotypes of the descendants of two individuals in the following generation, they slipped into population reasoning at the level of allelic frequencies, while continuing to use the Punnett square, whose function they changed. They shifted to a population reasoning at the level of allelic frequencies while continuing to use the Punnett square, whose function they use.

They presented a theoretical situation of 80% black alleles and 20% white alleles, using deterministic reasoning to say that the urn did not vary from one generation to the next during reproduction. The teacher said:

> So, in the population that results from the population, I always have 20% of light alleles 80% of dark alleles made by the population that are in the table... At each reproduction I will keep the same proportion. If we had done the exercise with 50/50 chance, we would have found the same thing, always 50/50 in all generations.

However, the students seemed to consider that there could be variations in these proportions. One student said: "At some point, it must be 17% or 65%... Chance must fall more on one of the blacks than...." However, the question of chance was not discussed and the teacher remained with a deterministic presentation of the law of probability. Genetic drift could perhaps have been addressed by questioning the importance of the size of the population, but we believe that this deterministic representation seemed to impede and prevent any discussion on this subject.

This deterministic vision (it always gives the same results from one generation to the next) is found in the following excerpt where the teacher considered the mathematical model to be based on the fact that all butterflies reproduce and that all gametes produce descendants:

> Yes, [randomness] plays a role, it plays a role on several levels. You agree that, in order for me to show you that the population does not evolve, I considered that all butterflies produce gametes and that all gametes produce descendants. Is that how it is in real life?

Here, we can observe another shift linked to a deterministic view of the model: the teacher stated that "all gametes participate" (deterministic reasoning) rather than "all gametes have the same chance of participating" (probabilistic reasoning).

12.5. Discussion and conclusion

The epistemological analysis presented in this chapter is focused on the concept of randomness in evolutionary science and mathematics. It allows us to analyze the activity of a teacher and their students during an attempt to introduce a probabilistic model in initial work on the evolutionary concepts of mutation, selection and genetic drift in the ELS class at the end of secondary school. We have thus been able to identify a certain number of difficulties or conceptions that hinder the understanding of probabilities in the teaching and learning process of these concepts. In the case study presented here, the teacher tries to get the students to engage in probabilistic population reasoning, where they are expected to construct neo-Darwinian explanations of population evolution that are focused on the conditions that make possible the evolution of allelic frequency distributions within populations over successive generations, but the goal is far from being achieved.

Specifically in this chapter, we looked at how the teacher introduced a probabilistic model, the Hardy–Weinberg model, to understand the evolutionary forces at play in the evolution of a population. The teacher's intention here was to work on deviation from the model, with this deviation then allowing hypotheses to be made about the causes of evolution.

From an epistemological point of view, we therefore considered two hypotheses:

– on the one hand, the non-distinction between a sample space that cannot be modified in a mathematical situation and the variability intrinsic to living organisms, with a sample space that is constantly changing, could constitute a rather strong obstacle for the students, especially if it remains implicit;

– on the other hand, in the classical situations in probability approached in college, the different trials are always carried out from the same starting device, so there is a stability of the "experimental device", i.e. of the urn in which the draws are carried out. This is not the case in evolutionary biology where the urn of gametes varies from one generation to the next and where the historical perspective is essential.

This case study shows a barrier among students related to the common meanings of the term "dominance". It can refer to the idea of dominance versus recessiveness or to the idea of a higher frequency that "dominates" in the population. This is a word-obstacle that weighs heavily on students' predictions of population change.

The deterministic presentation of the model makes it difficult to explain the phenomena of genetic drift and population evolution. By presenting it in a deterministic manner, the teacher probably wanted to allow students to see more clearly the evolutionary causes other than those related to randomness. However, by doing so, it seems to us that they are preventing discussions about genetic drift.

The use of simulation software seems to us to be essential in modeling the evolution of populations while maintaining a non-deterministic probabilistic approach. As these tools simulate a large number of trials in a short period of time, they allow us to go beyond a discourse that is simply based on examples in a reasoning that is not theoretical and mathematical. Students can vary a number of parameters (selective values, population size, etc.). The other advantage of these tools is to present the evolution of a population without necessarily having access to the underlying mathematical models, which may be too difficult for the students at this point and are not essential for understanding the evolutionary concepts. Other solutions are also possible. For example, Masel (2012) proposes, drawing on the history of science, population genetics topics based on progressively increasing mathematical difficulties.

The articulation between mathematical concepts of probability and concepts from life sciences is not self-evident. In the situation studied, it is the students who must build bridges between what they have seen in probability in their mathematics class and the use of these concepts in the life sciences class. It seems important to us that this transfer should not be the responsibility of the students, especially since, as we have seen, there are important epistemological differences between the use of these concepts in mathematics and in life sciences. It therefore seems to us that it is essential for teachers of these different disciplines to work together and to make the epistemological differences explicit. We insist on the importance of training on the issues of linking the teaching of probability and that of evolution, especially since the new programs for the final year of high school put forward this work based on probabilistic models in evolutionary sciences, in particular the Hardy–Weinberg model (BO special edition No. 1 from January 22, 2019).

12.6. References

Banet, E. and Ayuso, G.E. (2003). Teaching of biological inheritance and evolution of living beings in secondary school. *International Journal of Science Education*, 25(3), 373–407.

Beatty, J. (1984). Chance and natural selection. *Philosophy of Science*, 51(2), 183–211.

Conry, Y. (1987). Darwin et Mendel dans la biologie contemporaine. *Darwin en perspective, Revue de synthèse, Vrin, republié*, 123–138.

Darwin, C. (1989). *L'origine des espèces au moyen de la sélection naturelle ou la lutte pour l'existence dans la nature*. Editions La Découverte, Paris (originally published in 1859).

Fagot-Largeault, A. (2009). L'évolutionnisme comme science historique. In *Darwin : 200 ans*, Prochiantz, A. (ed.). Odile Jacob, Collège de France, Paris.

Gayon, J. (1992). *Darwin et l'après-Darwin : une histoire de l'hypothèse de la sélection naturelle*. Kimé, Paris.

Gayon, J. (1994). Le hasard dans la théorie évolutionniste moderne : une analyse philosophique. *Bull. Hist. Episém. Sci. Vie.*, 1(1), 4–15

Gobert, J. (2020). Modèle historico-épistémologique du système conceptuel aléa-sélection-hérédité-population, pour les enseignements et apprentissages d'une pensée populationnelle probabiliste évolutionniste en classe de sciences de la vie et de la Terre. *Recherches en didactique des sciences et des technologies*, 22, 239–263.

Gobert, J. and Lhoste, Y. (2021). Conquérir un point de vue populationnel et historique pour apprendre l'évolution, analyse de la co-activité professeur-élève par l'étude de la structuration des contextes. *Education & didactique*. Accepted.

Gould, S.-J. (2002). *The Structure of Evolutionary Théory*. The Belknap of Harvard University Press, Cambridge, MA and London.[2]

Goux, J.M. (1979). Les modèles en génétique des populations. In *Élaboration et justification des modèles*, Delattre, P. and Thellier, M. (eds). Maloines Edition, Paris.

Hodge, M.-J.-S. (1987). Natural selection as a causal, empirical, and probabilistic theory. In *The Probabilistic Revolution*, Kruger, L. (ed.). The MIT Press, Cambridge, MA.

Huneman, P. (2009). Sélection. In *Les mondes darwiniens*, Heams, T., Hunemann, P., Lecointre, G. (eds). Silberstein, Syllepse, Paris.

Jacob, F. (1970). *La logique du vivant*. Gallimard, Paris.

Jacob, F. (1981). *Le jeu des possibles*. Librairie Fayard, Paris.

Laplace, P.-S. (1986). *Essai philosophique sur les probabilités*. Bourgois, Paris (originally published 1814).

Le Guyader, H. (2012). *Penser l'évolution*. Éditions Actes Sud, Arles.

Malaterre, C. and Merlin, F. (2009). L'(in)déterminisme de l 'évolution naturelle : quelles origines pour le caractère stochastique de la théorie de l'évolution ? In *Les mondes darwiniens*, Heams, T., Hunemann, P., Lecointre, G., Silberstein, M. (eds). Syllepse, Paris.

Martin, V. and Theis, L. (2016). L'articulation des perspectives fréquentielle et théorique dans l'enseignement des probabilités : regard sur un changement de posture chez un enseignant du primaire. *Revue canadienne de l'enseignement des sciences, des mathématiques et des technologies*, 16(4), 345–358.

Masel, J. (2012). Rethinking Hardy–Weinberg and genetic drift in undergraduate biology. *BioEssays*, 34(8), 701–710.

2 See Gould, S.-J. (2006). *La structure de la théorie de l'évolution*, Paris, Gallimard, for the French edition.

Mendel, G. (1961). Recherche sur divers hybrides végétaux. *Bulletin de l'union des Naturalistes de l'Enseignement public*, 3, July–September (French translation).

Morange, M. (2005). *Les secrets du vivant : contre la pensée unique en biologie*. Odile Jacob, Paris.

Orange Ravachol, D. (2012). *Didactique des sciences de la vie et de la Terre. Entre phénomènes et évènements*. Presses universitaires de Rennes, Rennes.

Orange Ravachol, D. and Beorchia, F. (2011). Principes structurants et construction de savoirs en sciences de la vie et de la Terre. *Éducation & didactique*, 5(11), 7–28.

Parzysz, B. (2019). Des fréquences aux probabilités : apprendre à modéliser. In *Enseigner les premiers concepts de probabilités. Un monde de possibilités !*, Martin, V., Thibault, M., Theis, L. (eds). Presses de l'Université du Québec, Quebec.

Petit, C. and Zuckerkandle, E. (1976). *Génétique des populations. Evolution moléculaire*. Hermann, Paris.

Rumelhard, G. (1980). La "fraude" de Mendel. *Biologie-Géologie (APBG)*, 4, 863–854.

Rumelhard, G. (1986). *La génétique et ses représentations dans l'enseignement*. Peter Lang, Bern.

Sober, E. (1984). *The Nature of Selection*. MAMIT Press, Cambridge, MA.

Theis, L., Mai Huy, K., Parent, A. (2019). Une situation-problème probabiliste pour travailler des conceptions erronées à travers une articulation des approches fréquentielle et théorique en sixième année du primaire. In *Enseigner les premiers concepts de probabilités. Un monde de possibilités !*, Martin, V., Thibault, M., Theis, L. (eds). Presses de l'Université du Québec, Quebec.

Walsh, D.M., Lewins, T., Ariew, A. (2002). The trials of life: Natural selection and random drift. *Philosophy of Science*, 69(3), 429–446.

Academic programs

BO Special edition No. 1 of January 22 2019.

Conclusion

Corinne FORTIN[1] and Julie GOBERT[2]
[1] LDAR Université Paris-Cité, Université de Paris-Est-Créteil
[2] ADEF, INSPE Aix-Marseille Université, France

At the end of this book, it seems that the panorama of current research in the didactics of evolution in the French-speaking world echoes that conducted in the English-speaking world. Indeed, the concerns about the acceptability of evolution among students, the difficulties encountered in teaching or the issues of university training for teachers are all similar.

However, this book also shows specificities concerning the importance of the nature of the knowledge involved. In particular, the complex relationship between the assumption of scientific knowledge and its implementation in teaching or training is explored and analyzed here. Hence, we can speak of the didactics of evolution insofar as the research targets the links between scientific content and the purposes of learning or teacher training, as well as the links between pedagogical scenarios and the mobilization of evolutionary concepts by the different actors, i.e. the students and the teachers.

Beyond providing an up-to-date overview, and even though new research opportunities remain to be explored, this book also offers perspectives in terms of conceptual research, curricular changes and instruments for evaluating the understanding of evolutionary concepts.

Finally, to facilitate the visibility of the different themes, the contributions have been grouped into three parts. It goes without saying that each of these parts

The Evolution of Life: Teaching, Learning and Training,
coordinated by Corinne FORTIN and Julie GOBERT. © ISTE Ltd 2023.

maintains a permanent dialogue on the challenges of research in the didactics of evolution in order to consider or reconsider the questions of teaching, learning and training in their cognitive, historical–epistemological and sociocultural dimensions.

List of Authors

Marco BARROCA-PACCARD
LirEDD
Haute École Pédagogique Vaud
Switzerland

Jean-Christophe DE BISEAU
Université Libre de Bruxelles
Brussels
Belgium

Jérémy CASTÉRA
ADEF
INSPE Aix-Marseille Université
France

Hanaa CHALAK
CREN
Nantes Université
France

Magali COUPAUD
ADEF
INSPE Aix-Marseille Université
France

Patricia CRÉPIN-OBERT
LDAR Université Paris-Cité
Université de Paris-Est-Créteil
France

Alice DELSERIEYS
ADEF
INSPE Aix-Marseille Université
France

Corinne FORTIN
LDAR Université Paris-Cité
Université de Paris-Est-Créteil
France

Magali FUCHS-GALLEZOT
Didasco
Université Paris-Saclay
France

Julie GOBERT
ADEF
INSPE Aix-Marseille Université
France

Yann LHOSTE
Lab-E3D Université de Bordeaux
Université des Antilles
France
and
Université Libre de Bruxelles
Belgium

Christian ORANGE
Université Libre de Bruxelles
Brussels
Belgium

Denise ORANGE RAVACHOL
CIREL
Université de Lille
France

Fabienne PAULIN
S2HEP
Université de Lyon 1
France

Jean-François PONCELET
Université Libre de Bruxelles
Brussels
Belgium

Francis ROUQUET
CREN
Nantes Université
France

Laurent THEIS
Université de Sherbrooke
Canada

José Luis WOLFS
Université Libre de Bruxelles
Brussels
Belgium

Index

A, B

C, D

E, F

G, H, I

M, N

P, R

S, T, W

Printed and bound by CPI Group (UK) Ltd, Croydon, CR0 4YY

13/09/2026

14960998-0005